Alternative Routes to the Sustainable City

Alternative Routes to the Sustainable City

Austin, Curitiba, and Frankfurt

Steven A. Moore

LEXINGTON BOOKS

A division of

ROWMAN & LITTLEFIELD PUBLISHERS, INC.

Lanham • Boulder • New York • Toronto • Plymouth, UK

LEXINGTON BOOKS

A division of Rowman & Littlefield Publishers, Inc.
A wholly owned subsidiary of The Rowman & Littlefield Publishing Group, Inc.
4501 Forbes Boulevard, Suite 200
Lanham, MD 20706

Estover Road
Plymouth PL6 7PY
United Kingdom

British Cataloging in Publication Information Available

The hardback edition of this book was previously cataloged by the Library of Congress as follows:

Moore, Steven A., 1945–
Alternative routes to the sustainable city : Austin, Curitiba, and Frankfurt / Steven A. Moore
p. cm.
Includes bibliographical references and index.
1. Urban ecology—Case studies. 2. Sustainable development—Case studies. 3. Austin (Tex.)—Environmental conditions. 4. Curitiba (Brazil)—Environmental conditions. 5. Frankfurt am Main (Germany)—Environmental conditions. I. Title.
HT241.M67 2006
307.76—dc22

2006023016

ISBN-13: 978-0-7391-1533-6 (cloth : alk. paper)
ISBN-10: 0-7391-1533-2 (cloth : alk. paper)
ISBN-13: 978-0-7391-1534-3 (pbk. : alk. paper)
ISBN-10: 0-7391-1534-0 (pbk. : alk. paper)

Printed in the United States of America

∞™ The paper used in this publication meets the minimum requirements of American National Standard for Information Sciences—Permanence of Paper for Printed Library Materials, ANSI/NISO Z39.48-1992.

For Alexander Brian Canizaro, Makenna Claire Moore,
Andrea Louise Canizaro, and Kaitlin Ainsley Moore

Contents

List of Figures

List of Tables

PREFACE

Not all books require a preface because the text itself discloses all matters of concern. In this case, however, there are three concerns that inform what follows: the first is my increasing uneasiness with philosophical proposals that would guide design practice, the second is the need to clarify at the outset who is speaking in the case studies that follow chapter 1, and the third is my own partial perspective of the cases.

Architects and planners may certainly benefit, as will any citizen, from reading philosophical works. After completing this work, however, I am convinced that designers do not require a philosophy to design well. On the basis of my previous work, this reasoning is somewhat surprising and in direct opposition to that of the 1980s and 1990s that witnessed, for example, the influential liaison between the architect Peter Eisenmann and philosopher Jacque Derrida. Not only Eisenmann but also a host of other well-regarded designers of that period employed a priori theories of various origins to guide their design decision making on a day-to-day basis. In retrospect, we find it rather easy to find fault with some of this work because it served the appropriated philosophy perhaps better than the *client*, however we choose to define this term. Philosophical word games cannot, I now think, be successfully translated into tectonic forms because they are incommensurable phenomena. Words give expression to what is in our minds, as do buildings, but buildings must also interpret the material conditions of the world so as to solve real social and environmental problems. Accepting this reasoning suggests that we should be skeptical, not only of Derrida's version of poststructuralism but also of any a priori philosophy, including that of sustainability, as criteria for good design.

There is, however, a problem with such reasoning because to argue that all a priori theories pose a threat to good design is itself an a priori theory. Such circular reasoning is only one example of the difficulty that pragmatists, in particular, often create for themselves. In becoming skeptical of foundational beliefs it becomes very easy to contradict oneself by making skepticism the shaky foundation upon which subsequent reasoning depends—thus my uneasiness about making grand philosophical proposals to guide design practice. It will be better to begin this book, then, with the notion that sustainability and pragmatism both are less coherent philosophies than discourses, or sets of tools with plural philosophical origins. To argue that designers do not need an a priori philosophy to design well is quite different from arguing that philosophy can be a helpful tool for anyone engaged in the a posteriori interpretation of buildings and cities. This is the approach in what follows and that is further considered in chapter 1.

Second, the reader should be aware that I have used many, but not exclusively, pseudonyms to identify the characters who tell the stories I have reconstructed in chapters 2 through 4. The public figures who have left a written record of their involvement in city making are identified by their real names because it would be difficult to conceal public identities in any event. Private figures, however, are given pseudonyms to protect their privacy unless they have specifically consented to being identified. The offering of confidentiality to respondents allows them, I am sure, to offer interpretations of reality that are far more reliable. How this circumstance fits into the book's research design is further considered in the appendix.

Third, it will not be difficult for the reader to ascertain that I live in Austin, Texas—one of the three cities studied. Although I had lived in the city for a short two years before beginning this project, my continued engagement with its people and places has certainly provided me with a perspective somewhat different than for Curitiba and Frankfurt. Some may argue that I am alternately too close to Austin and not yet close enough to the other cases to give a balanced and objective accounting of them. I will leave this to the reader to judge, but simply hold that living in a place is not quite the same as studying it. This is not an argument that privileges the objective view of the scientist. It is, in contrast, an argument that privileges the competing views of many local citizens. The distinction will become clearer in the pages that follow.

In sum, it is fair to say that all analysis takes place within peculiar circumstances of which the well-scrubbed laboratory is but one example. The task I have set in this analysis is to try to make sense of the topic at hand without scrubbing away the circumstances that make it rich.

ACKNOWLEDGMENTS

Writing books is often perceived as a solitary affair, and there are indeed dark moments when that characterization feels all too accurate. But any scholar will surely tell a different story, one filled with helpful colleagues, friends, institutions, and students too numerous to mention. This book must also acknowledge the remarkable generosity of the local experts and citizens who helped me to understand the places they call home. I was continually amazed by the openness and goodwill of individuals I met in each of the three cities I studied—they not only gave up considerable time on my behalf but also demonstrated genuine concern in telling their stories fairly and clearly. It is to them that I owe the greatest debt.

The academy is a particular kind of democracy, one that generally recognizes the responsibilities, as well as the freedoms, of citizenship. For example, one University of Texas colleague on whom I came to depend, Kurt Weyland, I have never even met. In spite of our purely instrumental connection, he spent considerable effort in correcting my early misconceptions of Brazilian politics. Colleagues in the science and technology studies program at Rensselaer Polytechnic Institute, particularly David Hess, were more than generous in responding to material that I presented there in the spring of 2003. I must acknowledge the similar intellectual gifts made by colleagues at various meetings of the Society for Philosophy and Technology, where I presented segments of this research—particularly Andrew Light and Paul Thompson. Vincent Canizaro, between clever chess moves, demanded that I explain how this book was other than just another list of best practices. My good colleagues at the University of Texas—Michael Benedikt, Kent Butler, Elizabeth Mueller, Michael Oden, and Sergio Palleroni—provided ongoing support by reading partial drafts over a period of six years. And without the assistance of translators Ralf Brand (who also coauthored from this research an article in the *Journal of Architecture*), Amanda Bothwell, and Fernanda Mulho Pereira I never could have understood the background history of Curitiba and Frankfurt, let alone the nuance of what was said to me by the citizens of those cities.

But books are not solely constructed in the academy. Friends are exploited as good listeners and occasionally provide sanctuary from both the world under investigation and the academy that stimulates inquiry in the first place. Without the sanctuary provided by Gerry Leader and Lucy Aptekar, this book might still be a rather large pile of unordered 3 × 5 cards.

In addition to helpful colleagues, friends, and inquisitive students, institutions have become indispensable supporters of scholarly books. Two successive Mike Hogg Urban Scholars Grants provided research assistance that framed the initial inquiry, and the Graham Foundation generously made possible subsequent field research in Frankfurt. A semester-long research leave from the University of Texas made it possible for me to concentrate on the piles of collected data long enough to make sense of so many seemingly conflicted stories. And finally, a University Cooperative Society Subvention Grant awarded by the University of Texas at Austin helped to make publication of so many words and images possible.

In all, this book, like the world it interprets, is very much a social rather than an individual construction.

ONE

A Tale of Three Cities

Until 1999, when I first traveled to Brazil, I routinely argued that sustainable urban development could be achieved only in a democratic society. This claim derived from a general familiarity with the literature and a personal predisposition to progressive politics. Although the conceptual link between sustainability and democracy is not universally held, it is commonly associated with *Our Common Future*, better know as the Brundtland Report, after its editor, the former prime minister of Norway, Gro Harlem Brundtland (WCED 1987), which has enjoyed significant global influence and is generally accepted as the seminal text on the topic of sustainable development. As a supporter of the Brundtland Report in theory, I did not expect to find in Curitiba—a Brazilian city commonly held to be an exemplar of the concept—political conditions that dramatically challenged the implicit political disposition of the Brundtland Report's authors.

My experience in that city so altered my understanding of what sustainable development might "look like" and how it might "show up" in a place that I began to doubt my own understanding of the concept. Still, it took another two years for me to conceptualize a study that would test the relationship between sustainability and democracy and thus settle, one way or the other, the doubt that festered. This study derives, then, from my own need to reconstruct an understanding of the diverse and sometimes conflicting urban practices identified by others as "sustainable."

Four years later, I am prepared to argue that sustainability is a story line or plot—not a scientific condition or concept—that tends to show up first, or more powerfully, in cities where citizens have historically engaged in public talk related to local political, environmental, and technological conditions. Like any set of arguments, this one requires definition of the terms employed in it and considerably more backing before its implications can be appreciated. That is, of course, the purpose of what follows.

1.1 METHODOLOGY AND METHODS

Inquiry is driven as much by the kind of disturbed certainty disclosed earlier as by the unanswered questions posed by art, politics, or science. Unfortunately for the reader, disturbed assumptions generally remain the unspoken motivations of authors, if for no other reason than such disclosure tends to be tedious. To avoid being either secretive or tedious I will begin by disclosing that this investigation is influenced by a set of ideas that are associated with American pragmatism—the approach to problem solving that derives from Charles Sanders Peirce, William James, and John Dewey, among others. Although we commonly use the term "pragmatism" as meaning "whatever works" or "realistic," these philosophers had a better-developed thesis in mind. In their view the test of knowledge was not "truth" but "utility," which is to say that within the community being scientifically correct is less helpful than being successful. The emphasis here is not upon what we think as individuals distinct from our community but what we do as citizens of that community. Adopting this position as my own is a self-conscious choice that runs like a spine throughout this book. I have taken this position for three reasons, the principal one being that the literatures of sustainability and pragmatism both consider democracy to be a central issue. Because of this theoretical compatibility, pragmatism is likely, I think, to be a good tool with which to investigate the problem at hand.

The second reason for employing pragmatist tools in this investigation is that the past decade has witnessed the emergence of proposals for something like "green pragmatism" (McDonough 2002), a pragmatist architecture (Ockman 2000), "pragmatic environmentalism" (Light 1996), as well as a pragmatist approach to urban planning (Blanco 1994). These works suggest that architects, planners, landscape architects, and environmentalists are being attracted to conceptual tools that were forged when these disciplines were preoccupied with other theoretical possibilities. Employing the language of pragmatism to construct an approach to sustainable urban development provides an opportunity to test at least three related traditions of scholarship—pragmatism itself, science and technology studies (STS), and the urban design disciplines (planning, architecture, and landscape architecture)—to assess whether they offer a coherent logic of development. By "development" I generally mean not only the contemporary projects of "developers," architects, and planners, but also the qualitative development of society, technology, and the environment in terms that consider our evolutionary prospects.

The third reason I have employed pragmatist tools is because this tradition does not require foundational orthodoxy—in other words, it is a good umbrella for ideas that can contribute to a particular project yet may ultimately rest upon metaphysical assumptions that are not entirely consistent with my own (Light 1996, 135). Some readers will surely consider this instrumental logic to be a form of crass intellectual opportunism or simply sloppy scholarship, but it will, I maintain, present itself as a coherent approach as the analysis unfolds.

Although the design disciplines have certainly made substantial contributions to the study of sustainable development, they have too often been of the "how to" or "best practices" variety. By focusing any discussion of sustainability upon the relative efficiencies of various techniques, how-to books tend to reduce a complex topic to the seemingly simple matter of selecting the right tool for the job. I argue later that this is an overly deterministic approach because it assumes that technological choices are asocial—meaning that efficiency is the only criterion that guides our choices.

The emergence of STS after World War II has rendered naïve the once-popular notion that technologies emerge in society owing to logic inherent in artifacts themselves. As a branch of sociology and philosophy, STS investigates how and why particular technologies emerge in particular places at particular times in history. Most salient to this study, STS is a method of analysis that has been increasingly applied to the study of cities as a whole rather than to only the constituent technologies like the roads, bridges, and electrical power that allow them to function. An implicit assumption shared by STS urban analysts is that cities can be interpreted as "sociotechnical artifacts," or systems whose boundaries encircle not only machines but also the people who manage, operate, and benefit from them. David Brain (1994) was among the first to apply STS to the story of architecture. It was, however, Eduardo Aibar and Wiebe Bijker's study of the late-nineteenth-century city plan to expand Barcelona beyond its medieval walls was the first major study of this kind (Aibar and Bijker 1997, 238). More recently Bijker's work on the scale of cities has been extended by Barbara Allen (Allen 2003), Ralf Brand (Brand 2003, 2005), and Anique Hommels (Hommels 2005). In all, the STS approach to the study of how cities become themselves is a very good fit with the broader interests of American pragmatism. What follows is a bit like trying on a suite of ideas to determine their fitness for problem solving. It is also an opportunity to assess what the design disciplines might themselves contribute to philosophy and STS.

Each paradigm of inquiry—positivism, postpositivism, critical theory, constructivism, and STS—brings with it strategic research methods that are consistent with an attitude toward knowing. The responsibility of the analyst is to employ strategic methods of data collection and interpretation that are consistent with his or her foundational assumptions or worldview. This is to say that an analyst who sees the world through the lens of traditional science is not likely to employ long interviews as a research method because they do not apprehend data that can be interpreted "objectively." The same can be said of a constructivist, but in reverse. She or he would not employ quantitative statistical analysis as the only research method because statistics would not apprehend the type of data required for reliable interpretation of some social phenomena. Pragmatists, however, tend to prefer research methods that are action-oriented rather than those that are epistemologically determined. In other words, pragmatists are less interested in laying claim to the truth than in catalyzing action. That is the case here, and it suggests that I am less concerned with traditional concepts of "objectivity" or "subjectivity" than with fairness and activism.

This attitude should not suggest to the reader that questions of epistemology, or how we know what we know, are unimportant—far from it. It does suggest, however, that epistemology is an issue that is better considered, unlike physics, after surveying the situations in which the evidence is gathered rather than before, because the investigation itself shapes the method. Therefore, I have reserved a more abstract discussion of methodology for an appendix. My purpose in adopting this sequence is to privilege empirical analysis over the purely theoretical. I mean by this that the attitude toward knowing developed in chapters 5 through 7 derives from analysis of the cases studied in chapters 2 through 4 rather than extruding an interpretation of the cases through an a priori theory. As I confessed in the opening of this chapter, having once supported a particular theory of sustainable development with very little evidence to back it up, I am not anxious to repeat the same mistake.

In spite of this structure, a few comments concerning qualitative and quantitative methods will be helpful to the reader at the outset. The case studies presented in the first part of the book rely upon qualitative methods borrowed from the social sciences. The reason for doing so is that quantitative tools simply do not lend themselves to understanding the messiness of the life world. Although quantitative methods do identify demographic patterns that are extremely important to our understanding of cities that already lay some claim to developing sustainably, they do not help us understand how public talk concerned with sustainability shows up in one city rather than another.[1] If qualitative research methods tell "thick stories," then quantitative methods tell "thin stories" that are simply too reductive—they exclude those variables that are difficult or impossible to quantify. My own bias, then, is that high-risk, high-gain qualitative methods are the most productive at this stage of inquiry.

To reconstruct the thickness of these stories I have relied upon three methods of data collection and interpretation that check the reliability of each other: grounded theory (Strauss 1987, 221), long interviews (McCracken 1988, 269), and historical methods (Groat 2002, 275). Grounded theory requires that "urban" analysts suspend whatever preconception that may have brought them to the scene and allow an explanation of what is found to emerge from participation in city life and observation in an inductive manner. This is not to say that grounded theory is entirely an inductive method—only that it begins that way. Frequent deductive reflection is required to build hypotheses worth testing. Second, I conducted open-ended, long interviews with informed local citizens to help me understand the situation from their perspective. Initial interviews were arranged before I arrived in each city, and subsequent interviews were recommended by initial contacts in a "snowballing" sequence of people on all sides of the question until the information became redundant. These were amplified by subsequent correspondence and telephone interviews to fill in the gaps in my emergent understanding. Third, I checked my own observations and the recollections of respondents against local histories, public documents, and the professional literature. These three sources of data were analyzed individually and then collectively using the methods of "content analysis" (Lincoln 1985) to build the categories that I subsequently used to interpret all three cities in light of the philosophical literature generally associated with pragmatism and STS.

Of course, qualitative and quantitative methods of analysis are complementary in that the former is ideal for building hypotheses and the latter for testing them. This is why chapter 5 employs the quantitative methods of Geographic Information Systems (GIS). For the moment, it will suffice to say that my intent in these case studies is not to objectively explain what is "true" but to reconstruct and thus better understand the competing versions of reality perceived by the actors in these thick urban stories.

1.2 SUSTAINABILITY AND STORIES

The idea that we should live sustainably begins with the observation that we do not. This modern observation was first made by a relatively small but influential group of environmentalists and scientists in the 1960s (Ehrlich 1968) and is based upon their historical analysis of economic expansion, population growth, and resource consumption related to subsequent environmental degradation. Their analysis substantiated the shocking claim that if the rest of the world consumed resources at the same rate as do Western societies, then the Earth's ecosystem would soon become exhausted and unable to reproduce itself.

This primary claim concerning environmental unsustainability suggests a corollary that is made by a second, if smaller and less influential, group concerned with social equity and distributive justice. The secondary claim suggests that more, not less, economic growth is necessary if we are to meet the legitimate aspirations of the world's poor. This logic has not only political but also ecological consequences because the just distribution of economic growth would, these authors claim, reduce further environmental degradation caused by populations living below acceptable minimum standards of well-being. This is to argue that it is only human populations that live well above minimum standards of well-being that can afford the aspiration to develop sustainably. Simply put, if your children are hungry, you are not likely to hesitate to kill the last "X" or cut down the last "Y" to feed them. These combined claims require that we reassess not only the consequences of past economic growth in the West but also implications for the world's future.

Sustainability implies a story line because it begins with the knowledge claim that contemporary consumption habits cannot be sustained (Eckstein 2003, 62). This prognosis offers only three narrative possibilities in response—denial, regignation, or hope.[2]

Denial, the most popular response, implies that life will go on pretty much as usual (UNEP 2004, 267). Those who respond to contemporary conditions in this fashion see in the urgent call of environmentalists only another Jeremiah prophesying doom and gloom or, worse, the boy who cried "wolf" (AtKisson 1999; Prugh 2000). Here, however, I must distinguish between a prediction and a story. Predictions claim some higher order of knowledge, based upon scientific or divine origin presented as the final word on the subject. Understood in this way, predictions tend to shut down conversation because "he has spoken." You accept or reject his view of the future based upon your assessment of his authority. My point in distinguishing between prediction and story is to argue that environmentalists have been far too willing to predict and too unwilling to participate in storytelling. As a result, they have often been ignored as false prophets when their predictions prove to be wrong.

The second narrative possibility is resignation to the mass extinction of species, including humans. This is a rather grim Malthusian prospect adopted by only a few (Ehrlich 1968; Hardin 1968) and tends to lead to proposals for "cooperative coercion" (Rees 2004) managed by the enlightened minority on behalf of the rest of us. As we will see in chapter 3, this is reminiscent of the technocratic approach adopted in Curitiba—an approach that is very attractive to those who value efficiency but discount human intelligence.

The third narrative possibility is that we retain hope by imagining some collective action that might alter the path of history. This alternative should be understood as an intersubjective construction about what might become true in the future, not a truth claim in the present. This is to say that stories are not causal but trial balloons that test contingent yet plausible futures (Crilly 2000). If the balloon ascends too quickly or is overinflated, it will burst. If it is underinflated and fails to rise, it will languish on the surface of the present. It is this third and more hopeful narrative possibility associated with sustainable development. The problem at hand is to inflate possibilities that will rise with our aspirations.

Each of these three alternatives is itself the construction of a story replete with types of actors who have ideals and preferred behaviors and who communicate in specialized ways and in a particular time frame. Sustainability, then, is the social construction of a story line that provides a historical alternative to the prospect of environmental collapse (Dryzek 1997).

Of course, all societies construct stories about themselves. We do so not only to distinguish our tribe from others but also to explain to ourselves how our ancestors came to live in a particular place in a particular way. Nevertheless, such stories are not fixed. They are edited over time by new ideas that first appear as social practices, even if only discursive ones (Taylor 2004, 17). As conditions change, which they inevitably do, the "foundation narratives" of all societies are periodically rewritten (Nye 1997). Therefore, one way to understand the emergence of sustainability is not as a historically unique situation brought about by a singular case of overconsumption but as a periodic and necessary rewriting of the foundation narrative of Western society. As have others before it, our current era requires new story lines if history is to unfold in a trajectory we can accept on behalf of those (e.g., future generations) who are now unable to speak for themselves.

Table 1.1 provides a simple analysis of five story lines that have shaped human understanding of the world at different times in history: the heroic, religious, scientific, economic, and sustainable.[3] Although they are roughly chronological in order of appearance, none of them has ever entirely disappeared from view, nor is my list comprehensive. When new story lines appear, they simply share the available space with those that have preceded them and with others that are less prominent. The sustainability story is just one layer of history that best describes the dilemmas of our time.

Thinking about history as a succession of stories seems, at least at first, an outrageous notion. Is it possible that simply telling a new story will alter historical outcomes? To think so we would have to assert that the story to be told is a very powerful one indeed—so powerful that it will convince a majority of our fellow citizens that fundamental change of their ideals, behavior, heroes, modes of communication, and time frames is in their interest (Eckstein 2003). But, as we will see in the case studies in chapters 2 through 4, stories do not cause history. Rather, they reflect real capabilities for history making as much as they catalyze action. If stories are received as utopian fantasies, unrelated to daily life, characteristic behaviors, and plausible outcomes, they will be rejected by citizens (Holub 1992; Moore

Table 1.1. Characteristics of Alternative Story Lines

Kinds of story line	*Ideals*	*Behavior*	*Actors*	*Modes of communication*	*Attitude toward time*
Heroic	Excellence	Compete	Heroes	Legends	Immortal
Religion	Goodness	Obedience	Saints and prophets	Scripture and prayer	Eternity
Science	Truth	Experiment	Scientists	Logic	Time-lessness
Economy	Growth (quantity)	Maximize	Consumers and business	Images and numbers	Now
Sustainable	Develop (quality)	Optimize	Citizens	Feedback loops	Renewal

2001; Fischer 2003). What is of most interest is how stories of sustainable development come to be told and how they conscript others to retell the story to their peers and live differently.

The emerging discipline of mimetics, which derives most directly from the work of Richard Dawkins, proposes that human evolution is influenced not only by the laws of natural selection and genetic mutation but also by the passing along of ideas from one generation to the next (Dawkins 1976). Because ideas, or "memes," affect not only human events but also the physical environment, Dawkins argues that, over time, they also influence blind evolutionary choices. For example, if we understand the idea of "rugged individualism" to be a meme of the American West that has been passed from generation to generation for the past century and a half, then it is perfectly reasonable to argue in Darwinian terms that a cultural landscape has been produced by rugged individuals who will in turn influence the possible choices of future generations. My point here is that the stories and "foundation narratives" we tell to each other have more than passing interest—they contain or suppress evolutionary possibilities. It is not that human evolution takes place within such a short period of time but that the ecological impact of rugged individuals has been of geological proportion that in turn influences adaptive behavior and so on. An extension of this memetic logic would be to hold that it is not only the ideas that are passed between generations that have evolutionary impact but also cultural habits or practices. This logic, however, is not without problems, as we will see in chapter 6.

Sociologists refer to the type of investigation conducted, next as *discourse analysis*, a methodology that began with Michel Foucault's assertion that individuals are generally held prisoner by the discourse into which they are thrown. In other words, Foucault argues that the categories of language we learn—our memes—limit our ability to interpret the world by imposing "interpretive frames" upon our worldview that, in turn, edit reality into meaningful compositions by ignoring and erasing some available information. Although contemporary analysts have made very productive use of discourse analysis, many (especially pragmatists) challenge the notion that individuals are prisoners of a single story. Dewey, in particular, argued that through reflection individuals are able to gain critical perspective on the existence of several competing stories that vie for their allegiance (Dewey 1991; Dryzek 1997; Fischer 2003). This is precisely what Benjamin Barber is concerned with when he argues that "political talk is not *about* the world; it is talk that makes and remakes the world" (Barber 1984; emphasis in original). In his view it is "public talk" that allows citizens to receive and transform both the story lines and physical settings of their communities.

To ordinary citizens, of course, "critical discourses" can be incredibly boring or simply opaque. This fact is not lost on those who prefer to describe transforming story lines as "attractive" rather than "critical" and to state that, if the goal is to conscript supporters into one's own network, then the stories told must appear not only powerful but also attractive to diverse audiences (Brand 2005). We should, then, distinguish between the "city of feeling" and the "city of fact" (Barber 1984; Rotella 2003, 10) because it is feelings, not facts, that encourage citizens to invest in the coauthoring of stories about a sustainable future.

Defining Terms

The need to distinguish between "critical" and "attractive" discourses, or between "feelings" and "facts," as a source of citizen motivation reminds me that I am already beginning to use several terms related to storytelling interchangeably, a fact that will certainly muddle the investigation. For clarity, I propose that we should not think of sustainability as a *concept* or even as a *discourse*, as Dryzek (1997) and Fischer (2003) propose, but as the ends in view of much contemporary *public talk*. In the vocabulary I use later, the different kinds of public talk that characterize a place add up to a story line or a limited horizon of possibilities that is something like a loose plot. Distinguishing between each of these terms will be helpful before we move on to the stories themselves.

We generally think of a concept as a fixed idea authored in the past by an individual. The Heisenberg principle and social Darwinism are good examples—they refer to specific logics. The existence of a new concept, however, means nothing unless it is in circulation, that is, part of a conversation that challenges older concepts held by at least some people. When challenged, concepts are not typically revised, but they are routinely replaced by those that are considered more robust or up to date. In the context of this investigation, I too will set aside the term "concept" for the moment as not particularly helpful.

Unlike a fixed concept, a discourse is an ongoing conversation in which meaning develops in a hermeneutic fashion between its participants. Dryzek (1997, 8), for example, defines *discourse* as a "shared way of apprehending the world" because participants in a discourse come to share an emergent vocabulary or set of meanings that did not previously exist. But what distinguishes a *discourse* from a *conversation* is that the former tends to be formal while the latter is more informal and inclusive. Especially in contemporary society, the term *discourse* has taken on an elite connotation that tends to exclude ordinary citizens. The term *meme* proposed by Dawkins is even more obscure and exclusive. As a result, I will employ the term *conversation* or, better, *public talk* (Barber 1984, 19), because it suggests a more democratic and fluid exchange of ideas that suits my intentions here yet still requires participants to be *conversant* in the topic at hand.

There are, of course, both private and public conversations. In the context of this study, it is public conversation that is of most interest—meaning talk that concerns the public good as opposed to talk that concerns only the individual. The line between private and public talk is admittedly difficult to determine, as the case studies of Austin and Curitiba clearly demonstrate. It is this blurry line between the public and private that is, of course, a basic problem of political philosophy.

Although we may want public talk to be universally accessible, it is not always so. After all, not all kinds of talk interest every citizen, so we tend to join only those conversations to which we are attracted. This logic suggests that those of us who join a particular conversation speak in a peculiar way—baseball fans are a good example. They are participants in a conversation that is quasi-public. Ardent baseball fans in any city fluently cite statistics related to the batting and fielding averages of team members who are particularly important in the world of athletic competition. Although their words may be quite transparent to their fellow sports enthusiasts, their statistical citations are opaque to other citizens even though they ostensibly speak the same language.

The citizens of Boston are, for example, the authors of many public conversations, some Irish, some Portuguese, some Catholic, some Puritan, some about the city's "Big Dig," and some about baseball played in the unique conditions of Fenway Park. Although each of these distinct public conversations includes only some citizens, together they add up to public talk about the city's past and future that includes all citizens, whether they can talk baseball stats or not. Barber (1984, 177) defines "public talk" as that which "always involves listening as well as speaking, feeling as well as thinking, and acting as well as reflecting." In what follows I will employ this meaning but add to it the idea that public talk comprises many competing quasi-public conversations that vie for our attention and allegiance.

If public talk is a shared mode of interpreting the world in the present, then I will define *story lines* as something like a meta-conversation—a shared way of making sense of the past and speculating about what might become true in the future. Like a discourse, a story line is always unfinished. Unlike a discourse, however, its vocabulary and syntax exist in the service of action. We might agree upon the line or trajectory of action in a story, for example, yet still object to some terms of a particular discourse—its vocabulary. This is not to suggest that the nuance of language is unimportant, but that specialized vocabularies can create obstacles to understanding that are better avoided. Richard Rorty (1989, 286) observes that we tend to listen most carefully to those who talk like us, share our vocabulary, and apprehend the world as we do. The flip side of this observation is to recognize that we tend to ignore or, worse, suspect those who talk differently. The argument I am building here is that we too often miss seeing the forest for the trees—that the alien nature of some vocabularies (Ebonics, for one example) deflects our attention from the action embedded in the line of the story. A story line, then, is akin to the structure and direction of an ongoing conversation—it is a synthetic plot whose ending is always just coming into view. This is to hold that vocabularies are less important than actions, plots, or stories.

This proposal to clarify the meaning of terms, however, brings with it a few problems. Although there have been countless attempts to define or fix the meaning of sustainable development, none has yet succeeded. Fifteen years of science, action, and metaphysical speculation do not seem to have yet produced agreement of what the term means—nor is more time spent in this effort likely to succeed because the variables under investigation continue to change so rapidly (Guy and Moore 2005). The volatility of the situation does not, I maintain, make the term any less valuable to us. We have, after all, productively engaged in public and private talk about "truth" and "beauty" for millennia without having yet agreed upon a single meaning for either. Although it will certainly frustrate those who wish to fix meaning for the purposes of accounting and controlling resource flows, understanding sustainability as an evolving story line rather than a fixed concept requires that we be constantly attuned to changing conditions, or what systems theorists call *feedback loops*. This logic suggests that fixing the meaning of sustainable development as a concept is precisely the wrong thing to do, because it would require us to ignore changing environmental and social conditions. I am proposing, then, that the direction and meaning of the sustainability story lines will continue to shift with the conditions in which citizens participate.

As I have argued, story lines are forged by many public conversations, some of which are more important that others. Although I make no claim to being comprehensive, I did find in the cases studied that stories of sustainable development require three kinds of particularly significant public conversations: these concern the *political*, *environmental*, and *technological* issues that confront communities. Of course, the boundaries between these three categories are very porous—in practice urban choices inevitably have political, environmental and technological consequences that are difficult, if not impossible, to isolate. That is the point—although other conversations may influence how citizens of a particular city "apprehend the world"—baseball talk in Boston or art talk in New York, to cite two examples—the conversations related to politics, the environment, and technology tend to be most influential in shaping an approach to sustainable development. This is not to say that neither baseball nor art are important, only that they are generally less directly linked to forging public choices related to sustainable development. For this reason I will outline the general structure of public talk concerning political, environmental, and technological issues before turning to the case studies themselves.

1.3 POLITICAL DISPOSITIONS

I recounted earlier that my 1999 trip to Curitiba, Brazil, launched this investigation. Because the political talk of and in that city appeared to me at the time as distinctly undemocratic, it required that I consider whether there might, contrary to common assumptions, be alternative political routes to the sustainable city. To test that question I have selected three cities identified in the literature as aspiring to sustainable development: Austin, Texas; Curitiba, Brazil; and Frankfurt, Germany. The selection of cases from three separate continents should not be understood as an implicit argument in favor of globalization theory, that each of these cities is subject to precisely the same forces of development and, as a result, can be interpreted by the same universal measures. The differences between these cities are substantial indeed. Not only are they located on different continents but their citizens have embraced, or received, very different political traditions. This is to say that Austinites, Curitibanos, and Frankfurters receive and interpret global phenomena differently. The responsibility of the analyst is to interpret conditions from a meso- or intermediate perspective that appreciates both local discourses and global forces at work (Frampton 1983; Haraway 1995; Moore 2001).

This observation requires that I anticipate a critique that will, no doubt, be voiced by those readers who object to solving the "environmental crisis" via political talk. Their objection, based upon scientific grounds, is that such life-and-death matters cannot be left to laypersons and the indeterminacy of politics. What is needed, these experts hold, is decisive action based upon good science, not indecisive muddling based upon popular misconceptions. A brief example will suggest a response to this critique.

Authors of two influential books (AtKisson 1999; Prugh 2000) argue that sustainable development is a function of three variables. Both agree that, first, society cannot consume renewable resources faster than they can regenerate themselves and expect future generations to enjoy a comparable quality of life. Second, they agree that society cannot dump wastes into natural sinks faster than they are absorbed and expect ecosystem to continue functioning. But here agreement ends—where one expert argues the third criterion for sustainability must be to reinvest the remaining nonrenewable resources to create renewable substitutes, the other argues that the third criterion must be to protect other environmental services, such as the conversion of carbon dioxide into oxygen. My rationale for teasing out this difference is not to suggest that these experts are in radical disagreement about the nature of sustainability but to question how it is that, given limited economic resources, we should reasonably decide which criteria are correct or better. My point is that the disagreement between these experts is not a scientific one but a political one. This is to say that choices about which criteria for sustainability are best are social choices about how we want to live, not scientific choices about what is true or more efficient (Winner 1977; Feenberg 1999). This is not to argue that there is no scientific basis for sustainability, only that there are many ecologically functional ways for humans and nonhumans to live together. There is no external script to which science can turn to make such political choices—it is up to us.

Accepting the political nature of sustainability helps to explain why related public talk appears in some cities before, or more forcefully, than in others. The three cities studied were selected not only because they fostered vibrant public conversations about sustainability but also because each of these public conversations was evoked, or at least aided, by the existence of what I will refer to as a *regime of sustainability* (Moore 2001). I borrow this term from social science in two contexts. The first is to argue that all political systems have an interest in scientific knowledge that serves to legitimize political action (Elzinga 1993, 277). In this sense a *regime* is a set of institutional arrangements or decision-making procedures that attempt to control the expectations of citizens regarding the benefits that derive from scientific knowledge. Sustainability is, therefore, a public conversation that generates politically useful expectations about the future.

The second context in which social scientists employ the term *regime* is specific to urban planning and refers to the informal arrangements that tend to grow up between municipal officials and the real estate development community (Stone 1989, 253; Imbroscio 1997, 113; Lauria 1997, 136). The logic of regime theory holds that sustainability is a discourse that inevitably comes into conflict with the private interests served by municipal dependence upon real estate taxes. This is an issue of particular significance in Austin, if less so in Curitiba, and marginally if at all in Frankfurt because those cities are less dependent upon local property taxes to initiate environmental or other kinds of action. Nonetheless, the existence of regimes of sustainability in all three cities is both an effect and a cause of viable public conversations about sustainable development.

The existence of many regimes of sustainability does not, however, suggest that all share the same aspirations. No other claim could be further off the mark. Rather, the evidence gathered suggests exactly the opposite—that regimes of sustainability employ some of the same rhetoric but abide by dramatically different political assumptions in pursuit of different goals.

This claim can be understood better by employing Barber's (1984) genealogy of Western democracy as modified in table 1.2. In his analysis, the Scottish and French Enlightenments produced not one but three distinct political dispositions that flavor our public talk: *liberal anarchism*, *liberal realism*, and *liberal minimalism*.[4] Barber's fourth category, *strong democracy*, is his proposal for a historical alternative to those previously developed, all of which he finds inadequate.[5]

What chiefly distinguishes these political dispositions is their attitude toward conflict. Where liberal anarchist regimes tend to deny the existence of legitimate social conflict, liberal realists suppress conflict, liberal minimalists tolerate it, and strong democrats employ the existence of social conflict to transform conditions. My argument is that the three cities selected for study—Austin, Curitiba, and Frankfurt—are, respectively, exemplars of these three historical political dispositions or kinds of public talk. How they relate to Barber's fourth possibility, strong democracy, is a subject of analysis within the case studies. Thus, the politics of sustainability are far from monolithic. In my analysis of the selected cases I will use Barber's categories of political talk to compare how citizens see themselves in relation to each other.

1.4 ENVIRONMENTAL DISPOSITIONS

As I argued previously, *Our Common Future*, better known as the Brundtland Report, established the dominant definition of sustainable development. That now well-known definition proposes that "[s]ustainable development is development that meets the needs of the present without compromising the ability of future generations to meet their own needs" (WCED 1987, 245).

Table 1.2. The Four Dispositions of Liberal Democracy

	Liberal Anarchism	*Liberal Realism*	*Liberal Minimalism*	*Strong Democracy*
Generic response to conflict	Denial— Conflict emerges only because of the illegitimate authority of the state.	Repression— Conflict is dangero us because it disturbs the contract certainty upon which growth depends.	Toleration— Conflict is natural in a diverse society and stimulates individual creativity.	Transformation— Conflict is natural and provides both an opportunity and reason to transform social conditions.
What is valued	Rights— Individuals are perceived as having asocial origins and rights.	Wisdom— A few very wise individuals are required to administer social order.	Freedom— Atomized individuals must be free to pursue happiness within market conditions.	Participation— Politics is a way of living in which the overlapping interests of individuals are best satisfied.
Idealized citizen-architect	Howard Roark— The mythical hero of Ayn Rand's 1943 novel.	Jaime Lerner— Former mayor of Curitiba and governor of Parana, late twentieth century.	Ernst May— Planner/architect in Weimar-era Frankfurt.	Sam Mockbee— Late-twentieth-century progenitor of the "Rural Studio" in rural Alabama.
Selected city	Austin	Curitiba	Frankfurt	?

The Brundtland Report and this generally accepted definition of sustainable development were constructed by twenty individuals from twenty countries over a three-year period (1984–1987) under the auspices of the United Nations–sponsored World Council on Environment and Development (WCED). Under such cooperative conditions, it is certainly fair to say that the initial Brundtland Report was the product of a limited public conversation, the result of which was not so much a text as a common understanding, vague as it may be, of what sustainable development aspires to be.

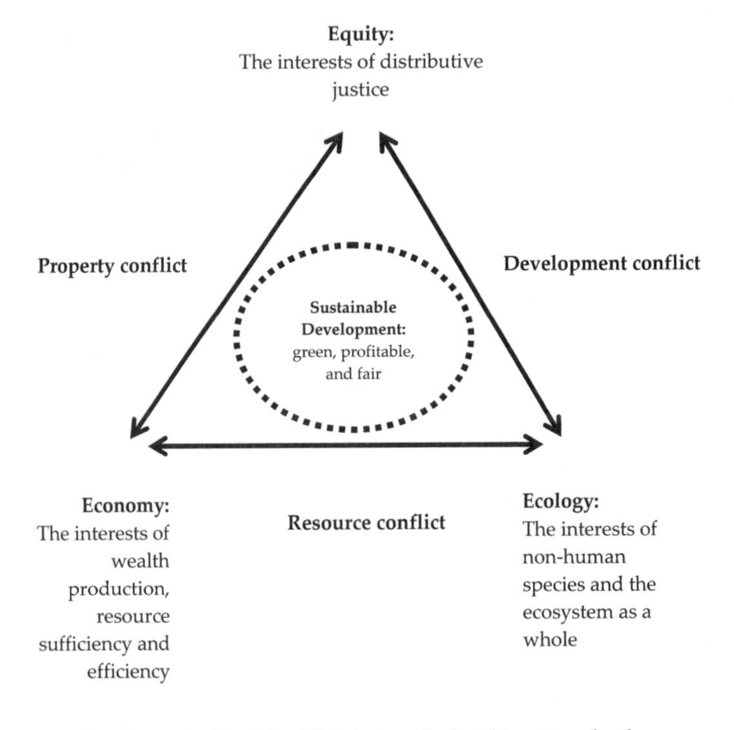

The Concept of Sustainability is inscribed within a triangle of competing interests. In this construction, the concept is necessarily discursive and democratic.
The Development Conflict sets those with an interest in protecting the environment against those with an interest in distributing available resources.
The Property Conflict sets those who control the means of production against those with an interest in distributive justice.
The Resource Conflict sets those with an interest in economic development against those with an interest in resource conservation.
The Sustainable City is one that negotiates and balances conflict and each set of competing interests.

Figure 1.1. The 3 Es

What emerged during this extended conversation was the understanding that sustainable development is achieved through the balancing of three competing interests within civil society: economic development, environmental protection, and social equity. This triangulated model—commonly referred to as the *3 Es*, or the *three-legged stool*—establishes a particular kind of relationship between the three variables from which we now expect sustainable development to spring up. The model, as formulated by Scott Campbell, is illustrated in modified fashion in figure 1.1 (Campbell 1996).

Campbell's model does an admirable job of illustrating the democratic interaction and discourse that the authors of the Brundtland Report foresaw as the route to sustainable development. The model suggests that achieving sustainability requires negotiating a balance between the competing social interests that alternately promote economic development, environmental protection, and social equity. The process of negotiating that balance requires public debate, compromise, and fairness. It was this fundamentally democratic logic that had led me to the conclusion in 1998, before traveling to Brazil, that democracy is a necessary, if insufficient, condition to achieve sustainable urban development. What I did not grasp at the time was that the logic of negotiated compromise implicit in the Brundtland model, and diagrammed by Campbell, reflects the assumptions of only one of the historical political dispositions documented by Barber (liberal minimalism) and only one of the many possible kinds of public talk about environmental issues that are documented by Dryzek (the Brundtland vocabulary). There are, however, many other discursive possibilities that derive from political traditions other than those of North America and Europe that might fit together coherently. We will encounter these other kinds of stories in the case studies.

On the basis of the case studies, my argument is that ecosystems and democracies are related and that both are complex, self-organizing systems in which highly ordered relationships can emerge without anyone consciously designing them. There are built-in negative feedback loops—such as floods, poverty, and contested elections—that can serve to move both ecological and social systems toward greater complexity and order. Although the Brundtland discourse illustrated by Campbell has clearly become the dominant kind of public talk associated with environmental issues, Dryzek has demonstrated, using the methods of discourse frame analysis introduced earlier, that there are other internally consistent logics in use that contradict those employed by the authors of the Brundtland Report (Dryzek 1997, 55). They are illustrated in table 1.2.

Thus, the possibility for arriving at a coherent set of public conversations that can be sustained over time appears to be far more complex than the authors of the Brundtland Report suspected. In his exhaustive review of environmental literature, Dryzek found that each of the existing public conversations about environmental problems could be categorized as being either

	Reformist	Radical
Prosaic	**Problem solving:** By experts Technocracy Administrative rationalism By citizens Pragmatism Lifestyle greens By the market Economic rationalism Ecomodernism	**Neo-Malthusianism:** Mutual coercion Survivalism
Imaginative	**The Brundtland vocabulary**	**Green radicalism:** Romantic Deep ecology Ecofeminism Bioregionalism Ecotheology Ecocommunalism Rational Social ecology

Figure 1.2. Environmental Dispositions

"reformist" or "radical" and as being either "prosaic" or "imaginative." By using these characteristics he could categorize kinds of environmental talk into four basic types, two of which have several subcategories, or variations upon a theme, as shown in figure 1.2. In my examination of the selected cases, it became clear that at least some citizens in each city subscribed to environmental talk other than that initiated by the authors of the Brundtland Report. This finding has the potential to create confusion about terms. To avoid that, hereafter I will use the term *sustainability* in a more generic sense that includes all environmental talk and use the term *Brundtland vocabulary* to refer to that initiated by the WCED.

In my analysis of the selected cases I will employ Dryzek's categories of environmental dispositions to compare how citizens described themselves in relation to nature.

1.5 TECHNOLOGICAL DISPOSITIONS

One way to understand the world's environmental crisis is that we suffer the unintended consequence of our own bad habits. For individuals or entire societies, habits are largely unconscious and repetitive patterns of behavior that may have life-enhancing, benign, or malignant consequences. Habitual exercise is certainly life enhancing, but being addicted to heroin or economic growth are habitual behaviors so malignant that they threaten our very existence (Booth 2004, 31).

It is commonly argued that technology is the most distinctive feature of modern society. Accepting this characterization suggests that human experience in the world is increasingly mediated by technologies of one kind or another. Talk about sustainable technology, then, not only is a matter of trading in one object for another but is about planning for an "alternative modernity" or an "alternative future" (Misa 2003, 271; Blanco 1994, 27). In the most radical of minds it is to suggest that modernity is not a condition determined by the adoption of particular technologies in a particular order but that there are multiple versions of modernity yet to be constructed (Taylor 2004, 266).

Choosing to understand unsustainability as an unnecessary and destructive social habit, as I do in this study, suggests that the reverse may also be true—that the condition of sustainability might spring from the conscious reconstruction of our social habits. The problem with this proposal is that convincing our fellow citizens to modify their habits is not enough in itself to alter the situation. This is because repetitious social behaviors do not exist in isolation from the built world. First, social habits coevolve with the technological systems that enable them. Second, once technological systems are in place they limit our choices to live otherwise (Hughes 1999, 273). Another way to argue this point is to say that the built world is the reification, or materialization, of our habits. If we are to consciously reconstruct malignant social habits, we must also reconstruct our technologies and landscapes (Latour 1987, 131; Rohracher 1999, 198; Moore 2001, 165; Brand 2003, 164; Misa 2003, 271).

If we use the term *technology* in the broad sense that includes not just technological objects like computers and automobiles, but also human practices and knowledge (MacKenzie 1999, 272), it is quite reasonable to claim that the very idea of sustainable urban development is a *technological story line* in the sense that it requires us to imagine how new technologies might transform landscapes constructed almost entirely by humans into functioning ecosystems. But this project will obviously be a political one, not a process of simply letting nature take over and burst through the concrete as so many romantic cartoons have depicted. We would have to decide what technologies will be used, at what cost, and under whose control they would be managed. In other words, it would be necessary to politicize technology in ways that we have not yet imagined (Hickman 2001, 97). Another way to argue this point is to say that "technology is not a destiny but a scene of struggle" that determines how we will live together and in relation to natural processes (Feenberg 2002, 15), or that conscious conversion to sustainable cities will be what Pfaffenberger refers to as a "technological drama"—a series of technological acts and counter-acts in which technological systems are regularized, adjusted, and reconstituted by a set of actors whom we implicitly recognize (Pfaffenberger 1992, 180). It will be helpful to the reader to think about each of the case studies in chapters 2 through 4 as urban technological dramas of just this sort.

If we want to understand how and why particular societies make the kind of technological choices they do, we need to understand the competing stories that are employed by local interpretive communities. My argument here, as earlier, is that in spite of such global conversations as surround genetically modified organisms (GMOs) or climate change, for example, interpretive frames are not interchangeable from place to place. Rather, interpretive frames are filters that are historically and spatially constructed by the public talk in particular places. The case studies will demonstrate and confirm this rather abstract logic.

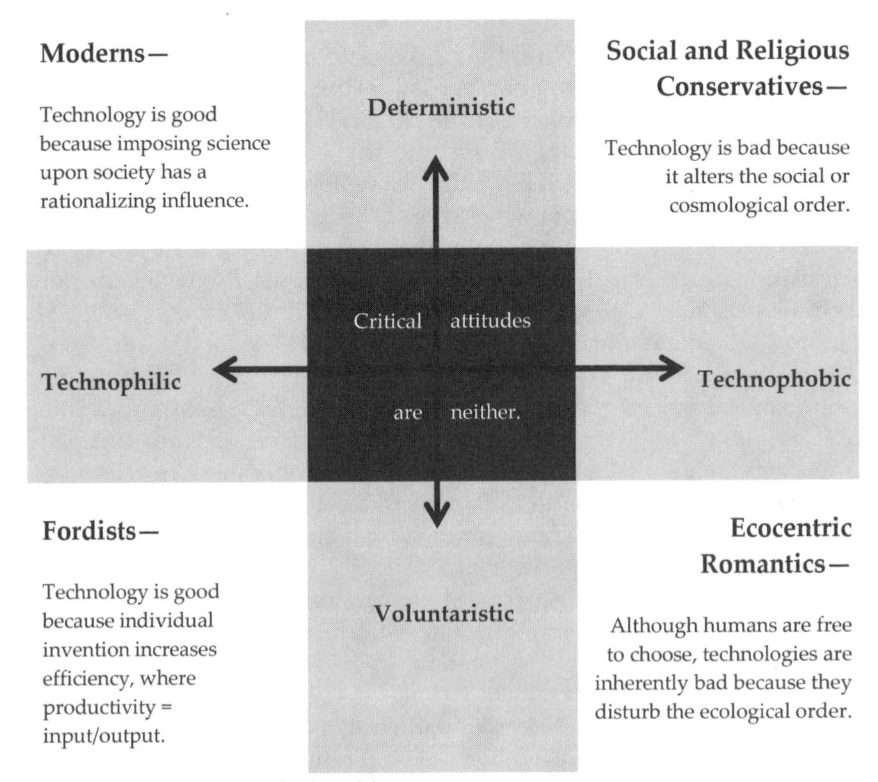

Figure 1.3. Technological Dispositions

In my examination of the selected cases I found that ordinary citizens in each city had interpretive frames, or attitudes, toward technology that tended toward the technophilic (meaning that technology is inherently good) or the technophobic (meaning that technology is inherently bad). Ordinary citizens also had split attitudes toward the relation of technology to society. Some were deterministic (meaning that they saw technologies as controlling society) and others were voluntaristic (meaning that they saw society as free to choose whatever technology was deemed desirable). These are what I will refer to as common interpretive frames and they are related to each other in figure 1.3.

More sophisticated observers—activists, STS scholars, and historians, for example—tend to have more critical views that are neither technophilic nor technophobic and neither deterministic nor voluntaristic. These various critical positions lie in the central cell of the figure. Because I have previously categorized critical technological public talk (Moore 2001, 165), as have others (Hess 1997, 268), I will not repeat that exercise here except by reference to specific authors.

As Barber has argued regarding political talk and Fischer has regarding environmental talk, the analysts' responsibility is to transform common attitudes into more critical or attractive ones. With regard to technology, this involves helping citizens see beyond the common discursive frames they have inherited but that do disservice to their own well-being. In my analysis of the selected cases I will employ these kinds of public talk to compare how ordinary citizens describe themselves in relation to technological change. I will also use technological talk derived from STS to suggest more critical interpretations.

To summarize this discussion of stories and sustainability, I will maintain that, although the idea that we can live sustainably is now a global phenomenon, it is not abstract models derived from deductive reasoning (for example, Campbell's illustration of the Brundland model) or universally applied lists of "best practices" derived from inductive reasoning (namely, LEED, the architectural checklist developed by the U.S. Green Building Council) that will catalyze successful action. These proposals certainly have heuristic value—meaning that they can introduce the naïve and skeptical to concrete concerns. Rather, my argument is that without reasonable talk models and lists are in themselves distractions from the social project of constructing future-oriented story lines from the situated perspective of locals.

The kind of future-oriented reasoning that I have in mind is what the American pragmatist Charles Sander Peirce referred to as "abduction," by which Peirce distinguished a type of reasoning that is neither deductive nor inductive but a way of pursuing "truth" that is developmental and self-correcting. According to Peirce, "Deduction proves that something *must be*;

Induction that something *actually is* operative; Abduction merely suggests that something *may be*" [emphasis in original] (cited in Magada-Ward 1999, 33). To say that something "may be" true is to say that we will never reach the final "Truth" of things, but that "truth" will improve over time—our understanding will get better as history (our collective experience) unfolds. It was certainly not lost on Peirce that his proposal was a radical one that would require reconstructing the "social and formal orders of inquiry" (Anderson 2006, 244). More than a century after Peirce's initial use of *abduction*, the reconstruction of inquiry is barely less controversial. Nonetheless, I am convinced that his insights provide a tool that is particularly well suited to the study of cities that aspire to develop sustainably because it captures the way that leaders in the cities I studied reasoned their way forward.

Following Peirce, the argument that unfolds next is that successful story lines of sustainability—meaning those that lead to satisfying action—are constituted of political, environmental, and technological talk that is homegrown from concrete ethical and environmental conditions. In what follows, I document how the most successful players in these stories employed abductive reasoning in solving problems on behalf of their communities. Before introducing the cases, however, a map of the inquiry will help the reader to anticipate the structure of the investigation.

1.6 THE QUESTION AND STRUCTURE OF THE INQUIRY

Are there alternative routes to the sustainable city? The authors of the Brundtland vocabulary imply that there is an a priori model of sustainability and that it is inherently democratic because it balances the interests of economy, ecology, and equity. Acceptance of this abstract logic suggests that there is a single path forward for those cities aspiring to develop sustainably. My initial investigation of Curitiba, however, suggests otherwise. But to argue that Brazil in general, or Curitiba in particular, is a nondemocratic society is fraught with difficulty and would, no doubt, offend many who live there. Nonetheless, Curitiba is not democratic in the sense that North American or European Union citizens expect. Neither, though, is the United States, following the presidential elections of 2000 and 2004, democratic in the way EU citizens have come to expect. It seems that the term *democratic* has, as Barber demonstrates, multiple meanings. His logic and my findings in the cases studied suggest that the hypothesis is different from that first posed. The question at hand turns not on the very existence of democracy as I first speculated but on the *kind* of democracy that exists in cities that aspire to develop sustainably. Citizens who have constructed different kinds of democracy will decide in their own way and on the basis of local conditions how conflicts are resolved and who gets to decide what technologies will be employed to solve which environmental problems. By differentiating between kinds of democracy we have the opportunity to determine whether one type or another better enables, or retards, sustainable urban development. Therefore, a better hypothesis with which to begin the investigation is that some forms of democracy are better than others in catalyzing sustainable urban development. The question to answer is "which?"

Toward that end, chapters 2, 3, and 4 are the case studies of Austin, Curitiba, and Frankfurt, respectively. Because the history and nature of these cities is so very different, so is the structure of each story. Each city unfolds in its own way—as it was told to me through interviews and as documented in local histories and public records. Within the analysis of each case my objective is to reconstruct the story lines that enable actions toward sustainable development and, conversely, to reconstruct those conditions and story lines that tended to frustrate successful action. Note, however, that I wrote "story lines" in the plural, because it turns out that there is inevitably more than one. There are dominant, counter, and even suppressed story lines in each city that demand our attention.

In contrast to the thick stories reconstructed in the case studies, chapter 5 relies upon quantitative methods. By employing GIS as a measurement tool, I provide different categories of analysis that will test the story lines constructed by local histories and citizens. Whereas the case studies interpret the three cities via historical and ethnographic methods, in this chapter I quantify spatial categories such as income distribution, amounts of open space per capita, and public transit routes. Another way to say this is that we need geography to discipline history and sociology. By analyzing six categories of space in each city, I find some surprising results, particularly in the case of Austin. Spatial analysis suggests that political stories do not evolve in isolation from spaces and technologies. They coevolve with them.

In chapter 6, I focus on the relationship between philosophical pragmatism and sustainable development. I do so by first reconstructing a set of twelve dilemmas derived from the stories of sustainable development told in each city. These can be characterized as alternative paths of action that appear, at least at first, to be equally attractive or unattractive. I argue, however, that a situated perspective of the problem at hand tends to suggest one path rather than another. The suggestion is not a generalizable fact or truth claim but a provisional experiment that requires future reflection, corrective action, and so on. My argument is that the cycles of action and reflection found in some successful aspects of the cases are entirely consistent with pragmatist attitudes toward the social construction of reality. I conclude, then, that pragmatism, sustainability, and urbanism do indeed share a *developmental attitude* that is highly complementary.

In arguing for the compatibility of these three discourses, I must emphasize that I am not arguing for anything like a pragmatist theory of architecture and urbanism. If we are to learn anything from the analysts associated with the pragmatist tradition it is that there are no foundational beliefs that can help us solve all problems in all places in all disciplines at all times. Rather, my interest here is to try out some ideas that have sprung up elsewhere and determine whether they are helpful in interpreting what happened in Austin, Curitiba, and Frankfurt. If so, the experience of the investigation should produce a few new insights of its own.

Chapter 7 concludes the investigation by arguing that there are indeed multiple routes to the sustainable city. The kind of democracy constructed along each route, however, may alternatively retard or enhance the chances of arriving at a condition of sustainable development. No city is fixed like the geology of moon rocks or stuck in the evolutionary stasis of alligator hide. Rather, we can understand all cities, as William James might propose, as dynamic cities-in-the-making. That each city must make its own way requires that we employ deductively reasoned models of sustainable development (even the one suggested by the Brundtland Report) and inductively reasoned lists of "best practices" (even LEED) only with caution. Finally, I argue that standardization of either kind will suppress, rather than enhance, the public talk required to construct the inclusive story lines that will help us to realize alternative futures.

Nevertheless, citizens require not only stories of their own making but tools. In lieu of models or lists I conclude by proposing twelve *abductive tools* that are derived from the dilemmas constructed in chapter 6. The purpose of these abductive tools is not to tell readers what to do to achieve sustainable development, and thus close public conversation, but to provoke new conversations about local possibilities. Readers more interested in these findings than in wading through the thick stories told by locals—on which the abductive tools depend—may simply want to skip to chapters 6 and 7.

A second warning, but to a different group of readers, is also warranted. Postmoderns who fear the return of master narratives—meaning those Enlightenment-era schemes designed to rationally predict and control nature on behalf of humans—may already be alarmed that sustainability is just the Enlightenment turned inside out: a scheme to predict and control humans rationally on behalf of endangered nature. Although this concern does have some merit, chapter 7 proposes something far more humble, yet still rational.

CHAPTER REFERENCES

Aibar, E. and Bijker, W. E. (1997). "Constructing a city: The Cerdá Plan for the extension of Barcelona." In *Science, Technology, and Human Values* 22 (1): 3–30.

Allen, B. L. (2003). *Uneasy alchemy: Citizens and experts in Louisiana's chemical corridor disputes*. Cambridge, MA: MIT Press.

Anderson, D. R. (2006). *Philosophy Americana: Making philosophy at home in American culture*. New York: Fordham University Press.

AtKisson, A. (1999). *Believing Cassandra: An optimist's look at a pessimist's world*. White River Junction, VT: Chelsea Green.

Barber, B. (1984). *Strong democracy: Participatory politics for a new age*. Berkeley: University of California Press.

Blanco, H. (1994). *How to think about social problems: American pragmatism and the idea of planning*. Westport, CT: Greenwood Press.

Booth, D. E. (2004). *Hooked on growth: Economic addiction and the environment*. New York: Rowman & Littlefield.

Brain, D. (1994). "Cultural production as 'society in the making': Architecture as an exemplar of the social construction of cultural artifacts." Pp. 191–220 in *The Sociology of culture*. D. Crane, ed. Malden, MA: Blackwell.

Brand, R. (2003). "Co-evolution toward sustainable development: Neither smart technologies nor heroic choices." Ph.D. diss., University of Texas at Austin.

Brand, R. (2005). *Synchronizing science and technology with human behavior*. London: Earthscan.

Campbell, S. (1996). "Green cities, growing cities, just cities: Urban planning and the contradictions of sustainable development." In *APA Journal* (Summer): 466–482.

Crilly, M. and A. M. (2000). "Sustainable urban management systems." Pp. 202–14 in *Achieving sustainable urban form*. K. E. B. Williams and Mike Jenks, eds. London: Spon.

Dawkins, R. (1976). *The selfish gene*. New York: Oxford University Press.

Dewey, J. (1991). *Logic: The theory of inquiry*. Carbondale, IL: Southern Illinois University Press.

Dryzek, J. S. (1997). *The politics of the Earth: Environmental discourses*. Oxford: Oxford University Press.

Eckstein, B. and J. Throgmorton, eds. (2003). *Story and sustainability*. Cambridge, MA: MIT Press.

Ehrlich, P. R. (1968). *The population bomb*. New York: Ballantine Books.

Elzinga, A. (1993). "Science as a continuation of politics by other means." Pp. 127–53 in *Controversial science*. T. Brant, S. Fuller, and W. Lynch, eds. Albany: State University of New York Press.

Feenberg, A. (1999). *Questioning technology*. London: Routledge.

Fischer, F. (2003). *Reframing public policy: Discursive politics and deliberative practices*. New York: Oxford University Press.

Frampton, K. (1983). "Toward a critical regionalism: Six points for an architecture of resistance." Pp. 16–30 in *The Anti-Aesthetic: Essays on Postmodern Culture*. H. Foster, ed. Seattle, WA: Bay Press.

Groat, L. and D. W. (2002) *Architectural research methods*. New York: Wiley.

Guy, S. and S. A. Moore, eds. (2005). *Sustainable architectures: Natures and cultures in Europe and North America*. London: Routledge/Spon.

Haraway, D. (1995). "Situated knowledge: The science question in feminism and the privilege of partial perspective." Pp. 175–94 in *Technology & the politics of knowledge*. A. Feenberg and A. Hannay, eds. Bloomington: Indiana University Press.

Hardin, G. (1968). "The tragedy of the commons." In *Science* (162): 1243–48.

Hess, D. J. (1997). *Science studies: An advanced introduction*. New York: New York University Press.

Hickman, Larry. (2001). *Philosophical tools for technological culture: Putting pragmatism to work*. Bloomington and Indianapolis: Indiana University Press.

Holub, R. (1992). *Crossing borders: Reception theory, poststructuralism, deconstruction*. Madison: University of Wisconsin Press.

Hommels, A. (2005). *Unbuilding cities: Obduracy in urban sociotechnical change*. Cambridge, MA: MIT Press.

Hughes, T. P. (1999). "Edison and electric light." Pp. 50–64 in *The social shaping of technology*. D. MacKenzie and J. Wajcman, eds. Philadelphia: Open University Press.

Imbroscio, D. L. (1997). *Reconstructing city politics: Alternative economic development and urban regimes*. Thousand Oaks, CA: Sage Publications.

Latour, Bruno. (1987). *Science in action*. Cambridge, MA: Harvard University Press.

Lauria, M. (1997). *Reconstructing urban regime theory: Regulating urban politics in a global economy*. Thousand Oaks, CA: Sage Publications.

Light, A. and E. Katz, eds. (1996). *Environmental pragmatism*. London: Routledge.

Lincoln, Y. and I. Guba. (1985). *Naturalistic inquiry*. Newbury Park, CA: Sage.

McCracken, Grant. (1988).*The long interview*. Qualitative Research Methods Series 13. Newbury Park, CA: Sage Publications.

McDonough, W. and M. Braungat. (2002). *Cradle to cradle: Remaking the way we make things*. New York: North Point Press.

MacKenzie, D. and J. Wajcman. (1999). *The social shaping of technology*. 2nd ed. Philadelphia: Open University Press.

Magada-Ward, M. (1999). "Rescuing Keller by abducting her: Toward a pragmatist feminist philosophy of science." In *The Journal of Speculative Philosophy* 13 (1): 19–38.

Misa, T., P. Brey, and A. Feenberg, eds. (2003). *Modernity and technology*. Cambridge, MA: MIT Press.

Moore, S. A. (2001). *Technology and place: Sustainable architecture and the blueprint farm*. Austin: University of Texas Press.

Nye, D. (1997). *Narratives and spaces: Technology and the construction of American culture*. Exeter: Exeter University Press.

Ockman, J. (2000). *The pragmatist imagination*. New York: Princeton Architectural Press.

Peirce, C. S. (1933–1958). *Collected papers of Charles Sanders Peirce*. C. Hartshorne and P. W., eds., Cambridge, MA: Harvard University Press.

Pfaffenberger, B. (1992). "Technological dramas." In *Science, Technology and Human Values* 17 (3): 282–312.

Portney, K. E. (2003). *Taking sustainable cities seriously: Economic development, the environment, and quality of life in American cities*. Cambridge, MA: MIT Press.

Prugh, T., R. Costanza, and H. Daly. (2000). *The local politics of global sustainability*. Washington, D.C.: Island Press.

Rees, W. (2004). "'Human sustainability' an oxymoron?" In the University of Texas Center for Sustainable Development, *Discussion Papers*. www.utcsd.org/ (accessed 12 July 2004).

Rohracher, H. (1999). "Sustainable construction of buildings: A socio-technical perspective." In *Proceedings of the International Summer Academy on Technology Studies: Technology Studies and Sustainability*, Inter-University Research Center for Technology, Work, and Culture, Graz, Austria.

Rorty, R. (1989). *Contingency, irony, and solidarity*. New York: Cambridge University Press.

Rotella, C. (2003). "The old neighborhood." Pp. 87–112 in *Story and sustainability*, B. Eckstein and R. Throgmorton, eds. Cambridge, MA: MIT Press.

Stone, C. N. (1989). *Regime politics: Governing Atlanta, 1946–1988*. Lawrence: University Press of Kansas.

Strauss, A. and J. Corbin. (1987). *Basics of qualitative research*. Newbury Park, CA: Sage.

Taylor, C. (2004). *Modern social imaginaries*. Durham, NC: Duke University Press.

UNEP (2004). *Annual report*. <www.unep.org/Documents.multilingual/Default.asp?DocumentID=67> (accessed 10 April 2005).

WCED (1987). *Our common future*. New York: United Nations World Council on Economic Development.

Winner, L. (1977). *Autonomous technology: Technics out-of-control as a theme in political thought*. Cambridge, MA: MIT Press.

NOTES

1. Portney's (2003) study of cities that "take sustainability seriously" is a good example of rigorously empirical study that documents indicators of cities already committed to sustainable development. See: Portney, K. E. (2003). *Taking sustainable cities seriously: Economic development, the environment, and quality of life in American cities*. Cambridge, MA: MIT Press.

2. The terms "alternative pathways," "alternative futures," and now "alternative routes" have become ubiquitous in discussions concerning sustainability but can be attributed to David J. Hess in the book series edited by him, *Alternative Pathways to Globalization*.

3. The terms and structure of table 1.1 are adapted from those by Betty Sue Flowers in a lecture at the University of Texas, 12 October 2000.

4. The terms and structure of table 1.2 are adapted from those by Benjamin Barber (1984), and appear in Moore amd Brand (2003).

5. The terms and structure of table 1.1 are adapted from those by Betty Sue Flowers in a lecture at the University of Texas, 12 October 2000.

6. The terms and structure of table 1.2 are adapted from those by Benjamin Barber (1984).

Two

The Springs of Austin

> If politics can be redefined as the public airing of private interests, public goods can be redefined as private assets.
>
> Benjamin Barber (Barber 1984)

In his investigation of American cities that "take sustainability seriously," Kent Portney identifies Austin, Texas, as one that ranks among the most serious (Portney 2003). And for good reason: the city has a Department of Neighborhood Planning and Zoning; it was among the earliest in creating a rigorous recycling program; its Green Building Program was the first of its kind in the nation and was subsequently honored at the 1998 Rio de Janiero environmental summit (ICLEI 2004) and served as a model for comparable programs in twenty-six other cities (Moore 2005); the city has perhaps the most successful sewage composting program in the nation; and the Central Texas Sustainability Indicators Project (CTSIP 2005), begun in 1999, is a model of regional introspection about what it means to be "sustainable." In the eyes of environmentalists from around the world, a great deal has been accomplished. (See figure 2.1.)

In Portney's quantitative analysis of twenty-four cities, Austin ranks in the middle for many categories of analysis but, nevertheless, qualifies for his more thorough study of eight cases. In Portnoy's analysis it is "surprising" that Austin measures up so well because the city's geographic location in the Sun Belt does not suggest association with such progressive ideas. This backhanded compliment suggests that the city has a particularly good, or at least interesting, story to tell. Its ranking in the middle of an elite group of cities that aspire to sustainability also recommends it as a case to study because it is neither the most serious (Seattle) nor the least serious (Milwaukee).[1]

Austin's story line unfolds in four parts. First, I tell the *received stories* that Austinites have heard from older citizens. These are from multiple historical sources, none of which is very controversial. They add up to a local history to which most would subscribe—it is the story of how Austin got "weird."

Figure 2.1. Location Map of Austin

The second part of Austin's story, which begins in the 1970s, is far more controversial. It chronicles what most refer to as the "water-quality wars" and the cycles of planning in which two competing groups—whom I refer to as rugged individualists and environmental preservationists—fought to describe the city's future. As we will see, planning in Austin can best be understood as four distinct campaigns that reflect cycles of economic boom, crisis management, and bust.

The third part of Austin's story examines the role that the city's form of governance has had in limiting the choices available to those who would bend the trajectory of history. Although preservationists had a powerful vision of a sustainable future for their city, their efforts were frustrated by the structure in which they had to operate.

In the concluding installment of telling Austin's story I reconstruct the two versions of reality reconstructed in this chapter—a dominant story line and a counter-story line. Yet, as we will see in chapter five, this seemingly dualistic story is not the final word. But for now, those who subscribe to the counter-story line of Austin were either not around when the dominant version was authored, not consulted, or explicitly excluded.

Figure 2.2. View of the Texas State Capitol from South Congress Street
Courtesy of the author.

These dominant and counter-story lines consist of various political, environmental, and technological public conversations that reflect a population deeply divided between the values of *individualism* and those of *environmental preservation.*

In all, Austin's story is weird, but not as weird as many would tell it.

2.1 RECEIVED STORIES

In light of Austin's colorful history as a Texas frontier town, its artistic reputation as the "world capital of live music," and its modern identity as a center for microtechnology innovation, Portney's surprise that Austinites take sustainability so seriously seems entirely reasonable. This unusual mix of identities requires some background if we are to understand the context in which sustainability sprang up as public talk among citizens. This background is best taken from local histories, various newspaper accounts, city Web sites, and interviews. Until the water-quality wars erupted in the 1970s, this received history was not a matter of contention.

Austin was not always "Austin." Originally Waterloo—a small settlement on the banks of the Colorado River in central Texas—the city was renamed to honor the Anglo-American pioneer Stephen F. Austin (1793–1836) when the new Republic of Texas wrested its lands from Mexico in 1839. Texas did not join the United States until 1850, and the ensuing Civil War conspired to keep Austin a minor post on the edge of civilization until 1871 when the first railroad arrived, followed ten years later by the election of Austin to become the seat of the University of Texas (Orum 1987; see figure 2.3). Thus, forty years after its initial founding Austin had secured its economic niche as a regional center for government and higher education. This blessing, or curse as some would have it, remained unchanged for nearly a century, until the mid-1970s.

The economic niche of the city seemed a curse to those who, like businessman Alexander P. Woolridge, wished for industrial expansion. Toward this end, in the 1880s Woolridge proposed the first dam to both control the unpredictable Colorado River and produce electricity for manufacturing. Completed in 1893, the dam spurred short-lived growth until it collapsed in the heavy rains of 1900. In spite of this setback, by 1920 the service industries of Austin grew to support nearly 35,000 citizens (Humphrey 2003).

Such growth and the nationwide city beautification movement of the day prompted civic leaders to engage in a professionally orchestrated planning process, the first since the city's founding, that resulted in the city council's adoption of the 1928 *Plan for Austin* (Sheldon 2000). As we

Figure 2.3. The University of Texas "Tower"
Courtesy of the author.

will see in chapter 5, this plan had long-term spatial implications that continue to haunt political and environmental conditions in the city. For the moment, however, we can understand these planners as they understood themselves—as advocates of growth, cosmopolitan beauty, and modern convenience.

Only a year after the adoption of the 1928 *Plan for Austin*, the Great Depression swept the world, leaving Texas as devastated as the rest of the country—except for Austin. Austin's curse, its lack of economic diversity and dependence on state government and higher education, proved to be a blessing in disguise because these services were relatively depression-proof. It was not only the structural economic conditions of the city, however, that saved it from economic and social despair. Elder statesman J. P. Buchanan and soon-to-be U.S. Senator Lyndon Johnson, close allies of the Roosevelt administration, were able to secure for the region the lion's share of Texas's New Deal funding for a variety of public works projects. Principal among these was a series of four new dams across the Colorado River that transformed its inconsistent and turbid flow into a chain of riverine lakes that not only served agricultural interests but also created the pristine hill country landscape that is so valued by contemporary inhabitants. (See figure 2.4.) In short, the Great Depression was an ironic godsend to Austin and cemented its political identity as a progressive

Figure 2.4. Tom Miller Dam
Courtesy of the author.

"New Deal town" (Orum 1987). In a state generally wracked by the "boom-and-bust" economic cycles of farming and oil production, the city's relative stability and progressive culture assured it an anomalous identity within the state.

Most southern towns before World War II were dominated by a traditional elite made up of wealthy landowners and a few businessmen. Unlike southern towns to the east, Austin created no wealth from agriculture because the land was unsuitable for that purpose. Nor did it produce much wealth from business, industry, or oil. As a result, the level of social equity in a population comprising bureaucrats and academics was unusual. This is not to suggest that some elites did not exist, but only that they were few and insecure in their hold on power. As a result, by the 1940s—an era of rampant populism and Leftist politics in pockets around the country—a new class of civic leaders showed up in the city. A succession of reformist city councilors, beginning with Emma Long (elected 1948) and Ben White (elected 1951), disrupted the political status quo in favor of less-affluent citizens. The move to the Left was certainly aided by citizens affiliated with the university and with the unions based in the capital (Orum 1987).

Affiliation with the Left was, of course, not universal. For example, a coalition of real estate interests was able to defeat the 1968 Fair Housing Ordinance that was designed to eliminate racial segregation in housing and promoted by progressives. They did so by playing "the race card" (Davidson 1990) and convincing homeowners that their property would become worthless if the ordinance passed. The example is instructive in that it demonstrates the significance of Texans' view of property rights even in the seemingly progressive island of Austin.

Well before this display of conservative power, business interests had begun to lay the foundation for a new approach to economic growth in Austin. In 1948, businessman and chairman of the chamber of commerce, C. B. Smith, arranged for an economic consultant from New York, Richard Wood, to develop recommendations to capitalize on the Works Progress Administration (WPA) investment in central Texas infrastructure. Wood's report recommended attracting research and development companies—"clean industries"—that would take advantage of the educated populace but would not compromise the quality of life and residential character of the city. Although this formula seems entirely predictable today, it was entirely novel to post–World War II businessmen in the American Southwest. That these men of commerce would recognize the economic value of the landscape they had helped to create suggests that they were early participants in public talk about the environment that John Dryzek refers to as "economic rationalism"—the idea that individuals, who are in competition with one another, can best serve society *and* nature by satisfying their own interests (Dryzek 1997).

Obviously dominant in the city council of 1955, the business community got the council to entrust creation of the *Austin Development Plan* to a subsidiary group of its own choosing. Adopted in 1961, this economic development plan provided incentives to invest in the city and successfully attracted IBM to the city in 1967, followed by Texas Instruments (1969), Motorola (1974), and Lockheed (1981). These influential microtechnology companies were subsequently followed by others so that by the mid-1980s Austin, unlike other Texas cities that were dominated by agricultural and oil interests, had succeeded in attracting a set of seemingly clean industries synergistic with the university and its fragile ecology. It turns out, however, that computer chip manufacturing is far less benign than assumed by early entrepreneurs, but viewed from Dallas, El Paso, Laredo, or Houston, in these years, Austin was just plain "weird"—an identity embraced by its residents.

After this point in Austin's history—the beginning of the high-tech boom—interpretations of history begin to diverge. In my interviews with representatives of the development community, developer Johnny Rae Morford in particular, I found a deeper understanding of the role that public investment by the Roosevelt administration played in the dramatic growth and political struggle for sustainable development that occurred after 1976. In Morford's interpretation of reality, the on-and-off economic boom of 1976–2000 was not caused by the arrival of microtechnology in Austin, but rather by microtechnology wealth acting on what I will characterize as the WPA landscape that already existed in central Texas. The construction of dams and infrastructure, as well as investment in the University of Texas by the WPA, created latent value that became available for appropriation by footloose capital. Observers in other locales have, like Morford, recognized that such "amenity environments" are attractive to capital precisely because they are "weird" and so differentiate the market. In the context of the neoclassical economic landscape in which Austin now sits, the purpose of public investment is not to create stable environmental and social communities but rather to create opportunities for the accumulation of private wealth. This is to say that, in the economic environment of the late twentieth century, amenities created by previous generations are understood as food in the unbroken chain of consumption. From this fecund background sprung a bitterly contested war about planning the city's future.

2.2 CYCLES OF PLANNING

In spite of, or perhaps because of, the protracted water-quality wars that took place in the city after 1976, Austin has achieved an international reputation as a city committed to the principles and practices of sustainable urban development. To locals, Austin's planning process seems far messier and conflicted than it does to those outsiders who watch from afar and consider Austin an example to follow. Of course, the urban planning process is the arena in which dominant and counter-story lines come into direct conflict. (We will see this pattern repeated in my studies of both Curitiba and Frankfurt.) In spite of the chaos experienced by some, however, Clark Allen argues that the Austin planning process can be characterized as having four distinct episodes. Mapping these planning campaigns against the cycles of the economy suggests that each period of planning was separated from the next by an economic downturn—a pattern that correlates the cycles of planning in Austin to those of the region's "boom-and-bust" economy (Thompson 2001).

Awareness Building

Allen refers to the first of these planning campaigns, 1976–1981, as a period of "awareness building" that took place in response to the first wave of in-migration and population growth. The political climate that characterized awareness building was, as in many cities that experience unanticipated growth, a simplistic battle between those who Allen describes as "rationalists and romantics or pro- and anti-growth forces." Kate Best, a veteran environmental activist, recalls that the formation of a water-quality task force in Austin was enabled by state legislation of the late 1960s designed to regulate non-point-source water pollution. The task force undertook the review of conditions and standards for each of Austin's twenty-five watersheds one at a time—an approach that earned Austin its initial reputation as being innovative in its attempts to preserve the environment.

What is striking about this effort, according to Allen, is that it was undertaken by a group of citizen-amateurs without any training or expertise in water management. Although advocates of participatory planning might hail this fact, his observation is that without the authority of science behind them, or their adversaries within the development community as complicit participants, the findings of environmentalist-citizens became immediately suspect as the ideological work of "no-growthers."

In this context we can associate Allen's critique of citizen-led efforts with the first definition of the term *regime* discussed in chapter 1—that science is routinely appropriated by politicians to legitimate action that is too often contrary to the knowledge cited. Being aware of the political value of scientific knowledge, he reasoned that to trump the interests of the developer-friendly city council, a new regime would have to base its proposals on "scientific objectivity" rather than political values that would not appear self-interested. Austin citizens, however, had by this time come to so distrust expert opinion regarding the environmental risks associated with development that they were more inclined to trust their own home-grown science than that of experts for hire. The tension or lack of trust between citizens and experts is an issue of general concern to pragmatists and to which I will return in chapter 6. For now, I will argue that the conflict in Austin adds credibility to the notion that the values of citizens with something to lose are no less rational than science—they are simply different types of rationality. At issue is our perspective on the consequences of change and the very idea that there is an objective view.

With or without the authority of science, the findings of the citizen-amateurs were widely distributed in the press, primarily by Daryl Slusher, then the environmental correspondent for the Leftist *Austin Chronicle* and later an influential city council member. Perhaps the most significant effort of this era was the first attempt at comprehensive planning since the 1928 *Plan for Austin*. Although the *Austin Tomorrow Plan* (1979–1980) was accepted by the city council, it never achieved the status of law because of the legal challenges made by the development community. The severe economic recession of 1981–1983 effectively halted the first planning campaign by slowing growth and temporarily defusing the crisis. But even without this time to reflect, many citizens of Austin recognized they had become, largely through their own efforts, very aware of the environmentally fragile context in which the city was making technological choices. This kind of public talk built awareness from the ground up.

Code Building

The second planning campaign, 1981–1990, might be understood as a period of code building. Recovery from the recession of 1981–1983 was enabled, in part, by the deregulation of the savings and loan (S&L) industry that in turn loosened lending regulations and encouraged massive development. As market forces were unleashed, the watershed-by-watershed process of regulating development by environmentalists culminated in 1986 with the passage of a comprehensive water-quality ordinance for the city. Parallel to development of the step-by-step water-quality ordinance was a second attempt by environmentalists to enact a comprehensive land-use plan. Although the drafting of the *AustinPlan*, as it was called, was for a short time considered a major victory by environmentalists, the most significant if unintended consequence of the combined water and land-use ordinances was, according to Allen, to galvanize the development community into action. Because the *AustinPlan* actually had legal teeth, it was taken seriously.

Speaking with a unified voice, the development community was able to convince the rural, small-town legislators of Texas, in collaboration with urban legislators from competing cities like Houston and Dallas, to pass House Bill 4 and Senate Bill 1704, which jointly declared it illegal for the city to legislate land use outside city boundaries but within its extraterritorial jurisdiction (ETJ). The combined bills were, as expected, quickly signed into law by then-governor George W. Bush. At the same time, the legislature acted to enable developers to subvert the city-owned water utility by forming their own municipal utility districts, or MUDs, and contracting with the Lower Colorado River Authority (LCRA) to actually provide services. Any delusion that the city once had of being able to control development by code building and through its monopoly of infrastructure was dashed. In the wake of the "Austin bashing" exercised by the Texas legislature, the *AustinPlan* was abandoned only two weeks before its final hearing. According to activist Sue Revanta, the city's planning staff was subsequently cut to a skeleton crew.

Revanta also reminds us that the deregulation of the S&Ls, the major source of development funding during the second planning campaign, came unhinged in the late 1980s through a series of dramatic bank failures and scandals that featured many of Austin's principal developers. In the months that followed the S&L scandals, Austin's environmental community became increasingly enraged. Passage of the federal Endangered Species Act at almost the same time served to fan the flames of popular awareness. The relevance of the simultaneity of these events is that the Endangered Species Act had obvious implications for the threatened Barton Springs salamander, a three- to four-inch-long blind creature that exists nowhere else on the planet. (See figure 2.5.)

Figure 2.5. Barton Springs Pool

Courtesy of the author.

On June 7, 1990, the prodevelopment city council was scheduled to take action on a proposal by Jim Bob Moffett, a developer from Louisiana much maligned by environmentalists, to develop a large parcel within the Barton Springs watershed. The public reaction, after smoldering for months—with considerable help from the newly formed Save Our Springs (SOS) Alliance—was an organic populist revolt. More than nine hundred citizens attended the hearing and signed up to speak against the proposal. Fearing the rage of citizens, the lame-duck mayor allowed the outpouring of citizen indignation to continue all night, until the framework for a new land-use ordinance, the so-called SOS Ordinance, was hammered out. Both Best and Allen agree that this citizen outburst should be understood in the larger context of California's populist tax revolt of 1978, Proposition 13, but with opposite political intentions. In Best's phrase, the city council and the chamber of commerce were given a "dramatic wake-up call." The economic downturn of 1990–1991, although not a full-blown recession, was sufficient to slow growth long enough for the players to catch their breath. Citizen reflection on the water-quality wars in this period can be characterized, in more ways than one, as a watershed event. Some simply became disillusioned because so much of their time and hard work had come to naught. These citizens dropped out, but others just grew angrier and dug in. For both groups the very idea of *planning* came to have a bad name.

When new cultural story lines come into existence—the story line of individual mobility, for example—they appear not only as technological artifacts (automobiles) and social practices (driving to work) but also as technological codes (manufacturing and traffic regulations). Andrew Feenberg argues that codes are the principal means by which societies formalize emergent social standards (Feenberg 1999). One way, then, to understand Austin's second planning campaign is as the struggle between two codes that would regulate differently how citizens would live in relation to nature. Without the ability to plan comprehensively, Austin's environmental community creatively tried, with some success and much frustration, to find other kinds of codes that might achieve the same ends. This was public talk about technology regulation.

The Regime of Sustainability

The third planning campaign, 1990–2000, was catalyzed by the emergence of the SOS Alliance, brilliantly named to give citizens a sense of their ownership of this threatened resource. In October 1991, however, the development community managed to get their allies in the city council to pass a revised water-quality ordinance that dramatically reduced provisions that citizens had insisted on during the famous all-night meeting.

Within hours of what Best characterized as the council's "betrayal," activists began an eventually successful petition drive to bring the water-quality ordinance directly to the people in the form of a referendum. In this game of cat and mouse, the prodevelopment council postponed the referendum from February to August 1992, allowing an additional 10,000 acres of land within the aquifer recharge zone to be platted and thus grandfathered from the provisions of the SOS Ordinance. When the referendum was finally held in August, the citizens of Austin voted two to one in favor of environmental protection (Austin 2006). In subsequent city council elections, the reformers ousted what John Logan (1987) has referred to as the "growth machine," meaning those council members associated with the development community, and got their own elected. As Best sees it, "Citizens took matters into their own hands" and reconstructed the political balance of power more in line with citizen sympathies. This slate of candidates, sometimes referred to locally as the "green machine," dominated city council politics until the spring of 2005 (Osborne 2003). It seems more than reasonable to refer to this new, if temporary, political order as Austin's regime of sustainability.

I have already made reference to the term *regime* regarding Austin in the context of how science is used to legitimate political action. The second use of the term by social scientists refers to the informal associations that tend to spring up between municipal officials and the local development community. The term is used appropriately in all three cases in both contexts, but the second context of its use is particularly relevant to Austin.

Students of *regime theory* commonly argue that local politicians are naturally attracted to those who control capital tied to "land-based" enterprises because such capital is relatively immobile, rather than "footloose," and has the greatest impact on city operations through property taxes (Imbroscio 1997). And, unlike in service industries, or even manufacturing enterprises—whose attention is drawn to national or international conditions—the principals of land-based enterprises have a natural interest and motivation to participate in local politics. The attraction works both ways and tends to contribute to what environmentalist Bill Bunch refers to as the good ol' boy style of doing business with "booze, beef, and blondes." This observation does not exactly make the informal affinity between local politicians and the development community a characteristic of transparent democracy. It does, however, make it a target for reform in a city that aspires to develop sustainability.

In hindsight, city administrator Michelle Conn argues that the SOS Ordinance forced a radical change in how the city's government responded to environmental issues. Academic observer Pat Roberts argues that the city stopped planning in any coherent sense and started regulating individual projects and technologies rather than space. But many citizens question the success of the ordinance itself, including SOS director Bill Bunch, because so many projects escaped its intent by being grandfathered. Twelve years after its passage, only a few projects have actually conformed to its requirements. Morford, turning the tables on environmentalists, has even argued that the ordinance has stimulated suburban sprawl because of its restrictive impervious cover requirements. But even Cam Stevens, the development community's lawyer of choice, recognizes that the ordinance has had significant "rhetorical success" in getting his clients and building contractors to accept other new technical standards and in getting major new industries like the Computer Services Corporation to locate in Austin because of increased contract certainty. Everyone knows the rules, yet no one will mistake them for a comprehensive land-use plan.

In the wake of the water-quality wars, attorney Kirk Watson ran for mayor in 1996 as a centrist against better-known candidates on the Left and Right. As an index of the city's cumulative exhaustion, he won the election with a mandate to bring the fractured community together, stop the bickering, and get on with the business of city building. Just as the new mayor got to work, much to his shock and amazement, the hated Senate Bill 1704, which restricted the city's ability to plan in its ETJ, was accidentally repealed by the Texas legislature in 1997. Although this story is in itself a book-length comedy, it must suffice to argue that Watson could not welcome this event as particularly good news because it threatened to begin the water-quality wars all over again. Anxious to display his ability to manage this unexpected crisis, Watson convened an emergency task force, with equal numbers of environmentalists and members of the development community, to work out an agreement that would take the place of SB 1704. The mayor's logic was that, if combatants could agree on new rules, action by the legislature would not be required when it reconvened eighteen months later (Spelman 2002).

According to former city council member Bill Spelman, the concept of "Smart Growth . . . was literally cobbled together in about twenty minutes so that members of the task force would have something positive to say to the press" following one of their first meetings. Smart Growth would become the set of new rules required to prevent both rekindling of the water-quality wars and new state intervention. The concept recognized the desirability of "growth" yet qualified how it would be achieved. As the negotiations wore on, each constituency at the table—environmentalists, the development community, and city staff—got through negotiated compromise something they wanted. After some months of work, city staff was able to construct a quantitative matrix that provided positive economic incentives for developers to pursue projects and specific plans deemed environmentally desirable by the city. Principal among the new rules was the delineation of a "desired development zone" in east Austin, on the other side of Interstate 35 and away from the delicate Edwards Aquifer recharge zone. Sue Revanta has argued that the *Smart Growth Initiative*, as it was officially named, was a brilliant stroke of imagination that enabled Mayor Watson to save a functioning, but threatened, political regime, and it was also an affective admission that the city had given up on comprehensive planning as a realistic goal.

The regime of sustainability had survived its first big threat and generally ushered in a new era of compromise. Real estate lawyer Cam Stevens, among others, was delighted by the improved climate under Mayor Watson. He still characterized Bill Bunch and his SOS Alliance colleagues as a generally "uncompromising" group of "environmental extremists," but also allowed that some of his own land-owning clients were no less extreme in their views about property rights. It had taken a few new voices, like that of entrepreneur and environmental activist Robin Rather, to get antagonists talking to each other and finding nonconfrontational solutions to mutual problems. One concrete result was the creation of the Hill Country Conservancy, a private nonprofit institution that worked with individual landowners to adopt conservation easements without City of Austin intervention. Even Morford likes to point out that the long battle over the status of the Circle C Ranch, a major suburban development within the city's ETJ in the hill country, was finally resolved in a spirit of compromise.

Although the implementation of the Smart Growth Initiative earned Mayor Watson a great deal of national and even international attention—and helped in his reelection campaign—he chose to run, unsuccessfully, for state office and never finished his second term. Some observers have argued that the regime of sustainability ended with Watson's unfinished term as mayor. Others hold that it was the "dot-com" economic bust of 2000 and the recession that followed that put an end to the regime of sustainability. As in the recessions of 1981–1983 and 1985–1987, economic downturns make any development look attractive to a property-tax-starved municipality. The Smart Growth doctrines were unanimously abandoned by the city council in June of 2003 in favor of what the local newspaper declared to be more developer-friendly rules (Osborne 2003).

Former councilman Spelman, however, takes a slightly different view—that "government tends to dilute reform movements by granting small victories." The small victories of reasonable compromise do preserve, even if in modified form, the specific interests of those who get to sit at the negotiating table. As architect and environmentalist Glen Braune recognizes, however, they do not fashion a collective vision of city life. His assessment suggests that the regime of sustainability, ushered in by the SOS revolt, was ended by the doctrines of Smart Growth itself. Like the triangulated model of sustainability discussed in chapter 1, the negotiated compromises of Smart Growth confined Austin within a Lockean universe in which citizens were atomized—never greater than the sum of their individual aspirations for the city or greater than individual economic interests. But as Austin *Chronicle* reporter Mike Clark-Madison has argued, "SOS will always have the built-in advantage of brand name, which is really the power of Barton Springs itself, as a symbol of how Austin feels not just about the environment but about democracy" (Clark-Madison 2002). This was public talk about democratic coalition building.

Regions and Neighborhoods

The fourth planning campaign, 2000 and after, can be called the era of regions and neighborhoods. Although many of the mechanisms of Smart Growth stayed in place after 2000, they did so without any of the ideological weight that put them there. As city administrator Michelle Conn argued in 2003, "Most environmental issues are regional, not local. So, a regional approach is required to really control garbage, watershed pollution, and especially air quality . . . because the city is bordering noncompliance with the federal Clean Air Act." Of course, the Texas legislature had not legalized land-use planning by county governments, let alone regional governments, prior to 2001, so as of this writing no test of a regional approach yet exists. Clark Allen argues, however, that all of the required mechanisms for countywide and regionwide planning were in place by 2004. The Envision Central Texas project (ECT) noted earlier in this chapter has been the principal tool of a regional approach. The successful referendum in support of a commuter rail plan in 2004 also added fuel to this fourth campaign, but it is too early to tell whether the purely voluntary nature of the ECT initiative will have the power to realize an alternative future. Without powerful incentives it is unlikely that many will see participation in the vision as being in their interest.

As the old ideal of regional planning was being reinvented in central Texas, there was a parallel effort to engage citizens directly in the planning of the city's sixty neighborhoods. Although the traditionally strong neighborhood associations applauded the turn to participatory planning, developer Morford noted, not without irony, that sixty neighborhood plans do not make a comprehensive plan. Allen agrees. In his assessment, neighborhood planning has been unsuccessful for two reasons. First, it focuses the attention of citizens on the interiors of their neighborhoods, not on the corridors that separate (and join) them to other neighborhoods. In affect, turning inward promotes NIMBYism—the "not-in-my-backyard" attitude that fosters reactive rather than proactive participation. Second, turning inward segregates neighborhoods rather than joining them in the larger project of building a common identity and citywide landscape. The cynically prone refer to this tactic of redescribing the city as so many atomized neighborhoods to be a "divide-and-conquer" strategy. By appeasing identity groups the regime could prevent, or at least inhibit, new citywide coalitions from forming. Members of the Austin Neighborhood Coalition (ANC)—a pan-Austin forum that exists to facilitate cross-neighborhood problem solving—will certainly disagree with this assessment. And lest we forget, Revanta argues that so much attention to the neighborhoods has effectively left the more vulnerable lands in the city's ETJ entirely unprotected. This was political talk about the relationship of parts to the whole. (See figure 2.6.)

In sum, my analysis of the four planning campaigns that took place after 1976 reconstructs the particularly colorful style of Texas politics but leaves the situation still unclear. Although it is true that a committed group of populist reformers (those affiliated with the story line written by environmental preservation) did manage to oust the traditional development-friendly municipal regime (those affiliated with the story line written by rugged individualists) and that the regime of sustainability did stay in power for a decade, the brand of sustainability that sprang up in the city seems superficial to those like planner Bob Paterson, who worked hardest to achieve it. The counter-story line of environmental preservation has not been powerful or attractive enough to overcome the structural conditions into which it was thrown. The dominant story line has successfully, at least so far, resisted the rewriting of fundamental institutional agreements such as property and water rights or regional cooperation between municipal governments that would make comprehensive

Figure 2.6. Typical Suburban Subdivision
Courtesy of the author.

planning and sustainable urban development possible. An analysis of the form of local governance will help to make this claim concrete.

2.3 FORMS OF GOVERNANCE

In this section, I argue that the form of governance adopted, or received, by citizens influences not only how decisions are made but also the decisions themselves. In the case of Austin, it is reasonable to characterize the city's social structure as pluralist—meaning, as I argued earlier, that there is no traditional elite that controls politics as is the case of many small southern towns. Rather, there are a number of competing groups—environmentalists, the high-tech community, academics, environmental justice interests, business, neighborhoods, no-growthers, and so forth—who must construct and reconstruct coalitions with other groups to get much of anything done. Among these competing interests, however, the development community comes closest to being dominant elites because they have the time and money to dominate public conversation. If we add to this generally pluralist profile the form of government that exists in Austin—which includes a weak mayor, a seven-member council who serve at-large (without geographic districts), and city manager—the difficulty becomes clearer.

The city manager form of government is particularly prevalent in the American South and is based on the assumption that the principles of scientific management can be applied to the problems of cities more effectively than can politics (Herson 1998). By turning over the management of infrastructure to experts, reformers of an earlier era hoped to rid technological and what they construed as purely administrative choices of partisanship—to make city governance scientific and cost effective. This form of governance, however, operates best in a small city or suburb that is relatively homogenous. In their analysis, Herson and Bolland (1998, 109) argue, "[H]eterogeneity leads to an overtly political format which in turn invites conflict, while homogeneity leads to a managerial format which in turn discourages conflict." Their observation recalls Barber's (1984) genealogy of political discourse that distinguishes liberal-anarchist regimes as those that deny the validity of political conflict. That Austin's liberal-anarchist regime would employ a city manager form of government appears, then, to be consistent with the political assumptions of the dominant group. My point here is that the simultaneous denial and discouragement of conflict leads only to suppressed anger, not to resolution or effective action. In the pragmatist perspective, public learning cannot occur without rational deliberation—without public talk.

It was in this context of conflict-denial that a diverse coalition of reformers tried to take control of the day-to-day technological and administrative choices of the city. To their frustration, however, it was a range of choices on which the city council had only limited control. City Manager Camille Barnett had effectively kept her staff isolated from political influence, but the city managers of later years, Jesus Garza in particular, were receptive to the environmental agenda of the reformers. Garza has even been accused of becoming the "yes-man" of Mayor Watson. Nonetheless, there remained institutional barriers that gave effective planning power to in-house technocrats and perennial underfunding to citizen initiatives of any kind. As Revanta reminds us—unlike the cities we will study later, Curitiba and Frankfurt—Austin received no continuous state or federal funding to plan or implement change, and there were no planning processes mandated by the state government as there were in other U.S. states such as Washington and Vermont. In a regional culture of conservative antitax sentiment and entrepreneurial expectation, the Austin city government had to pay its own way (Lauria 1997). As I noted at the beginning of this chapter, it is because of this larger political-economic context that Portney (2003) was surprised by Austin's willingness to take sustainability so seriously in the first place. To do so citizens have struggled with seemingly overwhelming obstacles.

One might argue that Austin Energy, the municipally owned electric utility, might provide an income stream to support environmentally sensitive innovation. Austin Energy is a remnant of the Roosevelt administration's New Deal plan to create universal rural electrification and is associated with the city's progressive history. The utility has adopted an aggressive and popular policy to generate renewable sources of power, primarily from wind farms in west Texas but also through solar installations. Like revenue derived from the city's sale of "Dillo-dirt" (composted sewage sludge), however, this revenue is plowed back into the general fund as a method of property tax relief. Rerouting this income to other environmental projects, difficult though it may be, would serve to politicize technology in a very productive way.

It is not particularly surprising, in light of its outmoded form of governance and perennial underfunding, that Clark Allen characterizes the planning process in Austin as a continuing exercise in "crisis management." Even his periodization of planning campaigns in Austin can be understood not as rising to periodic challenges but as a default admission that planning only occurs in the city when a major catastrophe seemed eminent. Prior to the SOS rebellion of 1990, Spelman reports that the condition of the city administration, and the planning department in particular, was so chaotic that no one even knew which projects had been approved and which had been denied. Nor could they find the drawings. Although the technocrats working under the city manager enjoyed some autonomy from raucous council politics, they had little accountability to either council members or citizens and few resources to actually do the job asked of them.

The argument here is that the city manager form of government is designed to depoliticize those very choices that should be at the center of public talk. Austin's form of governance may be fine for a small southern town or suburb with a homogeneous population, but it has been rendered obsolete by growth, diversity, and technological conflict. Former member of Congress Barbara Jordan (1936–1996), a major force in the progressive politics of her day, reportedly urged a change in Austin to a strong mayor form of governance. Modifying its form of governance would, of course, not single-handedly change Austin's style of conflict resolution. It would, however, make it possible for strong leaders such as Jordan to politicize individual technology and environmental issues and thus make them add up to a coherent story line that might guide city staff rather than allow staff experts to treat technological issues as asocial (Herson 1989).

At the risk of overemphasizing this point, problems associated with technocratic forms of governance also exist at the regional scale. The LCRA is the oldest and largest of ten such authorities in the state. Each river authority came to exist under separate legislation and is differently charged with the management of water resources outside city jurisdictions. Like the other river authorities, the LCRA is governed by a director who reports to a board appointed by the governor. These are political appointments that are generally made to individuals with ties to the development community. Simply put, this remarkably important institution effectively controls water policy in a semiarid region, and thus by default it controls rural development policy, with neither accountability to citizens nor any responsibility to coordinate their policies with regional city governments. In reality, the typical City of Austin and the LCRA have a twenty-year history of competition and conflict (Scheibal 2004). Or, as Revanta puts it, "The LCRA is like the organization that Robert Moses put together in New York City in the 1940s. . . . [I]t's hard to hold them accountable for anything."

Allen suggests that the problem is cultural as well as explicitly political. The positivist training and mentality of those who operate within this technocratic tradition prevents them from recognizing or taking responsibility for the social and environmental consequences external to their institutions' charter. This is a topic that we will return to in chapter 6 by distinguishing between what Frank Fischer refers to as "technological and cultural rationality" (Fischer 2000). That experts and citizens have come to think differently, and to assess environmental risk differently, is a problem that is exacerbated by Austin's form of governance.

The conflict between experts, citizens, and their political representatives raises interesting questions for Austin that will conclude this section. On the one hand, few American cities have witnessed a populist revolt of the scale or intensity seen in Austin. Although voter turnout at election time is rather average for the United States, the city is infamous in the region for the engagement of a core group of citizens in public life. On the other hand, however, are the structural barriers to citizen participation noted previously and the long history of "backroom deals" that exclude effective public participation in making choices about the environment and urban infrastructure. Two brief examples will help make the point. First, when Senate Bill 1704 was accidentally revoked in 1997 and newly elected Mayor Watson found his regime in crisis, he did not engage citizens in public talk about the future of their city. Rather, he secretly assembled a group of "experts" from competing factions to work out a deal. Watson had made it clear to his colleagues that he was not in office to get citizen input but to get things done. As a result, the citizens of east Austin, who woke up one morning in the city's "desired development zone," were enraged because they had not been consulted about the plans made by experts in secret about their neighborhood. This view of events, which I will call the "environmental justice story line," will be thoroughly examined later through the use of GIS. And second, as Bill Bunch reminds us, although the citizens of Austin have more than once voted overwhelmingly to protect environmentally sensitive areas in the Edwards Aquifer recharge zone, the Texas Transportation Commission has consciously employed Austin's tax revenue to build roads in exactly the areas citizens voted to protect. These examples, and more, demonstrate that citizen participation in Austin is both fact and myth.

Kate Best, the veteran environmental activist, perhaps unconsciously played with Thomas Jefferson's quotation concerning *freedom* when she argued that "the price of *democracy* is eternal vigilance." The type of vigilance that she seems to have in mind is the constant construction and reconstruction of inter-subjective stories—meaning those that share the hopes and aspirations of citizens. She holds that "without a powerful vision, council and staff can re-write the plan every week. What really counts is the day-to-day decisions that add up to a city."

Together these observations add up to an endorsement of a popular slogan, credited to Mike Wassinich, that began appearing on T–shirts, dumpsters, and briefcases across the city around 2000. "Keep Austin Weird" suggests not only a popular desire to recuperate the "old Austin" of pre-freeway days, but, as Michael Oden argues, to maintain the anomalous condition of relative equity and environmental integrity that has distinguished Austin from other Texas cities (Oden 2005). But this, of course, is Austin's counter-, not dominant, story line. With this background history, a summary of the city's planning efforts, and a critique of its form of governance as a foundation, I can now reconstruct both the dominant and counter-story lines that both document and drive Austin's history.

2.4 AUSTIN'S DOMINANT STORY LINE—RUGGED INDIVIDUALISM

> I came here to say that I do not recognize anyone's right to one minute of my life, nor to any part of my energy, nor to any achievement of mine, no matter who makes the claim, how large their number or how great their need. I wish to come here and say that I am a man who does not exist for others. (Rand 1943)[2]

Ayn Rand's fictional architect, Howard Roark, has enjoyed a long run as the hero of American individualism. Some, of course, employ him as a parody of this position because he is so overstated—so bigger than life, so . . . well . . . Texan. But the "rugged individualist" is not unique to Texas; he is a part of the American heroic era of settlement that extended from the eighteenth well into the late twentieth, if not the twenty-first, century and has influenced not only homesteading and ranching but business, science, and the arts. It is, however, a term rarely used as a complement anymore, except in conversations related to land settlement. As I have argued earlier, however, all kinds of public talk share space with competing political, environmental, and technological talk. For the sake of clarity I will reconstruct them one at a time.

Austin's Dominant Political Talk

Although few aspire to the stature of the rugged individualist, some, like Austin developer Johnny Rae Morford and real estate lawyer Campbell (Cam) Stevens, find resonance in the contemporary, if slightly less heroic era of suburban land settlement. Both men are commonly accused by environmentalists as being the guys in black hats—the self-interested actors who are responsible for the suburbanization of the fragile landscape to the west of the city. But, according to Morford, "We didn't bring people to Austin, we are only responding to market conditions." In the laissez-faire logic of Austin's development community—which includes not only developers such as Morford but also scores of builders, lawyers, bankers, and real estate brokers—cities are machines for the production of wealth. In a democratic society like ours, they argue, government exists only to facilitate private wealth accumulation. Or as Stevens puts it, "People want to live in Austin because it's the Berkeley of the Southwest and they have the right, if they have the money, to live in west Austin because its hills and lakes are beautiful." Both Morford and Stevens agree that there is no collective public greater than the sum of their clients' private property interests.

Those who see the world in this way privilege private property over public because of what Garrett Hardin has referred to as the "tragedy of the commons"—the notion that citizen number 1 has no incentive to save public resources because citizen number 2, number 3, or number 10 will surely buy another sheep to fatten on the grass that citizen number 1 left to regenerate the public pasture (Hardin 1968). In this story, a happy ending is assured only by privatizing everything and ensuring strong property rights that will optimize both the public good (meaning profitability) and environmental protection by avoiding the abuses typical of commonly owned resources.

Austinites Morford and Stevens understand property rights as reflecting the need for "predictability in life," or what philosophers would call *contract certainty*. Without certainty in the marketplace, economic growth cannot happen because the risk of investment would become too great. So, in the world of real estate development, when government changes the rules some of the potential value on which prior agreements were made may be lost. In Morford's asocial logic, if government changes the rules, for whatever reason, the public must be willing to compensate individuals for lost value. Otherwise, government would be exercising the power to take private property without compensation, which is in his view a constitutional contradiction. As city administrator Michelle Conn argues, for most families in the region real property is the primary source of wealth because its value appreciates, rather than depreciates, over time. As a result families of modest and substantial means are naturally resistant to government intervention. This natural reluctance is why the smaller cities around Austin—Dripping Springs in particular—have so vigorously resisted annexation into the city's ETJ. These citizens, according to Clark Allen, perceived annexation and land-use control by Austin's environmental community as a form of "taxation without representation"—what he characterized as a classical "libertarian David and Goliath story."

If Morford, Stevens, and the citizens of Dripping Springs employ libertarian rhetoric, which seems a fair characterization, they would seem to have much in common with the "liberal-anarchist" political discourse documented by Barber (1984). As I briefly noted in chapter 1, Barber argues that this political disposition is one of three that derive from the Scottish and French Enlightenments and each is characterized as having a different attitude toward conflict resolution in society. Citizens attracted to liberal-anarchism deny that any conflict should exist at all. Simply put, they deny any validity to the conflict created by Austin environmentalists because government, especially that of another city, has no right to tell homeowners and ranchers how to manage their lives and private property. People, rugged individualists argue, live in rural communities because they need enough room to move around. They do not want, or need, city people telling them what to do. According to Michael Badnarik, the Libertarian Party candidate for president in 2004, "the federal government's role should be limited to national defense, coining money, and operating the post office. The government . . . should not have a Department of Education, the Food and Drug Administration, or the Environmental Protection Agency" (Associated Press 2004).

Liberal-anarchists recognize that it is no accident that Badnarik is a long-time resident of Austin, because that city is one base of his support. Nor are libertarian values limited to the Austin city limits in Texas. In 2003 Governor Rick Perry made similar arguments in defending his overuse of the veto in the state's biennial legislative session—"Texas," Perry argued in a classically colorful (if undisciplined manner), "already has too many laws," so he simply vetoed most of the legislature's work for that session (Perry 2000). Perry's logic, like that of the real estate developers who support him, is that laws of nature ensure the dominance of natural elites (meaning entrepreneurs like themselves) in guiding the business of the community. Barber's vehement objection to this logic is that it subverts the very idea of democracy by inserting an "independent ground" or set of imagined and self-serving "natural rights" that are designed to trump the open-endedness of genuine public talk (Barber 1984). Austin's dominant kind of political talk—liberal-anarchism—should, then, be understood as antithetical to the democratic values implicit in sustainable urban development as envisioned by the authors of the Brundtland Report.

Austin's Dominant Environmental Talk

Glen Braune takes the long view of the conflict over property rights and its environmental consequences. As a native Texan of several generations, he recognizes that his fellow citizens take property rights far more seriously than do citizens of eastern states. "But," he maintains, "you have to remember that we are only two generations from the frontier. The way we think about land is changing because Texans are adaptive—we are still adapting to urban life." Braune's generous characterization of Texans, and their willingness to change, might even be applied to Johnny Rae Morford.

Morford claims to recognize that "many people don't approve of our [company's] typical projects. I know that some of our work is junk and it takes away from society. So, bottom line, I'm part of the [sustainability] problem. But being a part of the problem gives me the right to be part of the solution." Morford now sees himself in the middle between extremist environmentalists like Bill Bunch and the "good ol' boys" that dominate the development community of which he is a recognized leader. The environmental talk that he and his allies articulate is what Dryzek (1997) generically calls "economic rationalism."

In Dryzek's characterization economic rationalists subscribe to four related attitudes. First, they believe that the world consists only of individual economic actors—a position consistent with the liberal-anarchist political talk discussed earlier. Second, economic rationalists see individuals as having only a competitive rather than cooperative relationship to one another. This is a worldview that lends itself particularly well to the real estate market and to conceptualizing land as a commodity that seeks its "best and highest use." Third, although individuals are adequately motivated by self-interest, the system also needs a few disinterested individuals to establish the rules and norms required to create contract certainty. This is, of course, the role of self-appointed elites like Governor Perry and his political appointee. And fourth, the basic metaphors employed by economic rationalists tend to be mechanistic—meaning that individualists envision cities not as collective accomplishments but as machines for private wealth production. Alternately individualists tend to refer to rhetorical states of freedom, as in free markets or the free range.

There is, I must point out, more than a little irony in how the environmental attitudes of economic rationalists serve to misinterpret, or obscure, the origins of ecological degradation in the central Texas landscape that surrounds Austin. There is ample documentation that the initial degradation of this ecologically fragile landscape was the result of overgrazing. If we follow the logic of economic rationalists (and libertarians) who subscribe to the doctrine of "the tragedy of the commons" (Hardin 1968), one would expect to find a spatial correlation between overgrazed lands and those that are communally managed, like the *ejidos*[3] of the Hispanic communities that inhabit the American Southwest and Mexico. Likewise, one would expect to find a spatial correlation between profitable, well-managed lands and those that are privately managed. The irony is that exactly the opposite is closer to the truth. While the Hispanic, communally managed lands in New Mexico, in particular, were ecologically sustained for four hundred years, the economic competition between rugged individualistic ranchers in the Texas Hill Country led to serious environmental degradation, the rampant proliferation of nonnative species, and agricultural failure in only two or three generations (de Buys 1985). Austin's dominant environmental talk—economic rationalism—should, then, be understood as a convenient and self-serving myth.

Austin's dominant political and environmental talk adds up not only to attitudes but practices or ways of doing things. The fiction of the freedom of the range requires us, then, to consider dominant attitudes toward those technologies that enable an imagined life of freedom.

Austin's Dominant Technological Talk

Individualists see themselves as major actors in what David Nye calls the American technological discourse of "self-creation" (Nye 2003). A central theme in this kind of talk, as it manifests itself in Austin, is that homebuilders, bankers, and bulldozer operators, for example, perceive the first creation—God's work—as incomplete. What we as Americans have been called upon by God to do is finish the job—to engage in the second creation by recognizing the "pattern latent in the [land]" and applying the human technology required to realize that pattern (Nye 2003). In this view of the land there is no such thing as a bad technology because the new tools invented by gifted individuals invariably increase human productivity and thus our ability to perfect imperfect nature. In the eyes of individualists this is a creative act, not an act of exploitation. Allowing water to flow down the Colorado River into the Gulf of Mexico uncaptured is, for example, understood by rugged individualists as sheer waste because it is a potential source of wealth that literally passes them by. The building of dams to harness the God-given potential to irrigate crops and attract homebuyers is, however, humanness at its best, individualists argue, because it is harmonious with the latent pattern present in God's first creation. Some individuals are, the story line goes, able to sense the pattern more readily than others, and these are the great men who invent the technologies needed by society to move forward toward realization of the second creation.

The attitude of individualists toward history is certainly teleological—meaning that God's grand plan is written in stone, yet conveniently provides men with adequate flexibility to pursue their own happiness through technologies that are perceived as inevitable, but act within broad limits such as the Ten Commandments. Viewing the making of Austin through these lenses suggests that the railroad of 1871, the dams of the Depression era, the highways of the 1970s and 1980s, and the "dot-com" boom of the 1990s fulfill a grand scheme in the pursuit of individual happiness that was intended by God. Austin's dominant technological talk—self-creation—should, then, be understood as a story authored by a class of self-appointed voluntarists.

All three discourses are summarized as Austin's dominant story line in table 2.1.[4]

In summation, I will characterize Austin's dominant story line—rugged individualism—as a self-serving teleology that reflects the asocial worldview of American pioneers in Texas and elsewhere. There is a well-intended temptation, as articulated by Braune previously, to attribute such individualism to the harsh life endured by early settlers to the region. Accepting mass ecological degradation by employing the doctrines of environmental determinism is, however, a bit like blaming the victim for crimes committed against him or her. On these grounds I will argue that pioneering Texans need to recognize the ecological consequences of their old habits before the construction of new habits becomes possible.

Table 2.1. The Dominant Story Line of Austin—Rugged Individualism

Kinds of talk	As told by: Bill Spelman, Johnny Rae Morford, Cam Stevens, and Bill Bunch
Dominant Political Talk	*Liberal-Anarchism*
Generic response to conflict	Denial— Conflict exists only because of the illegitimate authority of the state
Attitude toward certainty	Cognitive— In pursuit of absolute freedom
What is valued	Rights— Without which autonomous individuals cannot realize their natural potential
Conceptualization of space	Infinite— Individuals, like Newtonian atoms, are naturally provided with adequate space to avoid social friction and maintain autonomy; social friction is caused by the state's illegitimate imposition of rules such as zoning
Dominant Environmental Talk	*Economic Rationalism*
Basic entities recognized or constructed	Homo economicus— Individual economic actors who seek resources
Assumptions about natural relationships	Competition— Individuals exist in a competitive relationship with one another and other species
Agents and their motivation	Self-interested individuals— In addition to individual economic actors, the system also requires a few disinterested individuals to establish rules required to maintain contract certainty
Key metaphors and rhetorical devices	Mechanistic— The social world is envisioned as a machine that produces desired products Free— as in markets and life on the range
Dominant Technological Talk	*Self-Creation*
Attitude toward technology	Technophilic and voluntaristic— Technology is good because individual invention increases human productivity
Attitude toward history	Teleological— God's grand plan allows men flexibility to pursue personal happiness through technology (within specific limits, namely, the Ten Commandments)
Assumptions about the origins of technology	Individual genius— Great men invent the machines that society needs
Types of tools employed	The most efficient— whatever works

But to be clear, my intention in this analysis is not to essentialize Texans as environmental rapists or explain why some individuals abuse the land more thoroughly than others. Such a claim would only obfuscate the situation because it would be equally true to argue that some pioneering Texas families constructed an ecologically viable landscape that served both humans and nonhumans well. My objective in this analysis is more limited—to understand the interpretive logic used by the dominant group and surmise how an alternative story line came to dominate the city, if only temporarily.

Austin's counter-narrative is no less Texan but emerged under political, environmental, and technological conditions that are generally suppressed from the contemporary view. In what follows I will use the same format as earlier to reconstruct this story line.

2.5. AUSTIN'S COUNTER-STORY LINE—ENVIRONMENTAL PRESERVATION

For those who have not lived in Texas and whose impression of the place is formed by exposure to old Western movies, presidential politics around the turn of the century, or the contemporary news media, it is altogether too easy to hold a stereotypical view limited to the conservative discourses I reconstructed previously. The reality of the place is significantly more complex, as are the characters who built it.

Austin's Political Counter-Talk

Many readers will reasonably assume that Texas has always been politically conservative—a state long dominated by the Republican values that gained national preeminence in the general elections of 2000 and 2004. For this group it will come as a surprise that in general elections of 1896 the Texas Populist Party officially received 44 percent of the popular vote—a percentage reduced by voter fraud, but still three times that of the Republican Party. Readers may be equally surprised to know that the state ranked ninth in the nation regarding the number of workers engaged in strikes and progressive political politics. In the context of late-nineteenth-century Texas politics, the Populist Party represented a diverse coalition of poor white Southern agrarians, Hispanics, and former slaves who organized themselves in east and south Texas, at least for a brief time, against the socially conservative and openly racist Democratic Party as well as the pro–big business Republican Party that dominated the Hill Country and west Texas.

It is commonly known that in this post–Civil War era social conservatives, unlike the progressive populists, registered to vote as what Southerners called "Yellow Dog" Democrats, rather than as Republicans, because they would sooner hang than join the party of Abraham Lincoln. This twist of history created a climate ripe for third-party politics. The Populist movement in Texas was eventually crushed by the divisive politics of race, brilliantly manipulated by both Yellow Dog Democrats and Republicans, but the history of progressive politics in the region is at least as long as that of land settlement itself. Although the Populist Party had itself become a bastion of "nativism" and white supremacy, by the mid-1920s the egalitarian legacy of the common man pitted against the new technocratic elite who ran the railroads, banks, and government became a permanent theme in the progressive politics in Texas. Thus, to argue that the progressive or political counter-talk of Texas is historically "populist" is to say that it is linked to the politics of the Southern yeoman freeholders who were generally Christian, egalitarian, and anticapitalist (Davidson 1990). Although this alliance of sectarian and progressive interests may seem very unlikely, today it is the backbone of Austin's political counter-talk.

The era in which the Populist Party emerged in Texas and elsewhere in America was, of course, the same as that in which Peirce and James developed the beginnings of pragmatist thought. This is hardly a coincidence and it is worth commenting that, where most philosophers were aghast at the rudeness of the populist program, pragmatists were sympathetic because at that moment in history both groups warned against what Louis Menand has called the "idolatry of ideas" (Menand 2001). Both Populists and pragmatists, still reeling from the absolute certainties that fueled the Civil War, were very reluctant to grant unquestioned authority to any claim. This epistemological skepticism was directed not only at the abolitionists who Southerners held responsible for the war but at the new class of experts noted earlier—those who controlled banking, the railroads, and government. Beginning in the nineteenth century, populists and pragmatists shared a skeptical attitude toward expert culture. This is particularly true where experts remained inflexible in the face of obvious suffering by common people.

In the post–World War II era of the civil rights movement, but before Yellow Dog Democrats left the Democratic Party en masse to become Reagan Republicans in the 1980s and 1990s, Texas progressives were led by a group of colorful figures who included Henry Gonzales, Ralph Yarborough, Lyndon Johnson, Barbara Jordan, Jim Hightower, Sissy Farenthald, Henry Cisneros, Anne Richards, and others. As late as 1976, progressives still exerted significant influence in crafting the state's public policy. And it was to Austin, of course, that progressives flocked from those remote parts of the state that became increasingly conservative, like the nation itself, after the shift in party identities that began in earnest during the Reagan era.

The move toward sustainable urban development in Austin during the final decades of the twentieth century must be understood in the context of the populist-inspired public conversations that emerged in the last decades of the nineteenth century. That the water-quality wars began in Austin at about the same time that poor whites deserted progressive Populist ideals, and their own economic interests, in favor of Republican ones is a highly significant piece of history. Without poor whites as race-bridging members of the political counter-talk, progressive-liberal values were left to the poor African American and Latino communities on the one side of economic security and to richer white intellectuals and environmentalists on the other. This is to say that progressive political talk in Austin became increasingly fragmented after 1976 and, as a result, ineffective.

Although nineteenth-century populists certainly saw themselves as continuing the yeoman democratic traditions envisioned by the founders, there has always been a *demagogic* strain in the character of many of the movement's leading actors. I mean by this term that larger-than-life personalities tended to accumulate power, presumably on behalf of common folk, through emotional personal appeal to the latent prejudices of citizens, rather than articulating a strong party platform or set of ideas. In this context populists have always been accused by cultural elites, sometimes legitimately, as being self-serving, undisciplined thinkers prone to inciting the rabble to thoughtless action. Although the Populist Party and its cousins, the Knights of Labor and the Greenback-Labor Party, have all disappeared from American politics, the populist tradition of personality-politics has not. In Texas, contemporary progressive activists who employ an evangelical style—former commissioner of agriculture Jim Hightower and the Jewish cowboy comedian Kinky Friedman[5] come to mind—operate in this tradition. My point is that the demagogic and intellectually undisciplined side of populism is antithetical to pragmatist proposals for public rationality and strong citizenship. In their view, demagogic populism has further weakened and isolated progressive public talk in Texas and elsewhere. On the positive side, however, colorful populists like Austin newspaper columnist John Kelso have kept alive the tradition of skepticism toward expert culture through a brand of political sarcasm that is truly Texan if no more disciplined.

Austin's political counter-talk—progressive populism—should, then, be understood as a historically significant attempt to construct common cause between intellectuals, poor whites, and communities of color. It has, however, been so thoroughly marginalized by its own demagogic tendencies and the neoconservative coalition of poor whites and big business that it is a caricature of its former self.

Austin's Environmental Counter-Talk

Planner Bob Paterson argues that the citizens of Austin have only just begun to understand the place where they live and that they have done it the hard way—by suffering the consequence of unanticipated natural cycles in a region prone to hills and dramatic, if infrequent, rainfall. City administrator Michelle Conn acknowledges the same conditions in saying that "the Glenrose and Edwards limestone that form the aquifer to the west of the Balcones fault are sponge-like formations that are highly responsive to rainfall events." This is a polite and technical way of saying that, when it rains twelve inches in an hour, as it does in central Texas from time to time, the very thin soil over the limestone makes for "flashy" events, meaning that it is not at all unusual to experience a twenty-foot-high wall of water charging down one of the city's generally dry creek beds. Historically, flash floods have taught immigrants to the region respect for the cycles of nature through tragic experience. In the Memorial Day flood of 1981, for example, twenty-five people died. In another example, in 1993 the city's central emergency headquarters was incapacitated by several feet of water—an ironic consequence of professional planning. The hard-learned lessons of some citizens have not, however, translated into development patterns that are consistent with the natural energy flows of the place. Experts argue, then, that planning decisions must not rely only upon improved knowledge, and that planning by experts alone is not enough. Paterson holds that "citizens must learn how to live here," they must consciously affiliate their development practices with the natural rhythms of the place.

Employing the same logic, environmentalist/lawyer Bill Bunch argues that Texas is in need of a "land ethic," a term commonly heard in local environmental discourse. This is a direct reference to the forester/author Aldo Leopold (1887–1948), who first coined the term in the *Sand County Almanac*. We can understand Leopold's proposal for a land ethic within the modern project of extending rights to an ever-wider circle of beings worthy of what the philosophers call "moral consideration." We can also understand it within the postmodern project of understanding humankind as a part of nature, rather than being distinct from it. In Leopold's own words:

> The land ethic simply enlarges the boundaries of the community to include soils, waters, plants, and animals, or collectively: the land.
>
> A land ethic of course cannot prevent the alteration, management, and use of . . . "resources," but it does affirm their right to continued existence, and, at least in spots, their continued existence in a natural state. (Leopold 1948)

The obstacle to realizing a Texas land ethic is, Bunch holds, that

> Texas has no public lands to speak of so citizens are detached from nature. In general, the Texas landscape is not spectacular so it takes more imagination [on the part of citizens] to get inspired. We are so big and so diverse that it's hard to construct an ethic from scratch. The collective idea of Texas is more like an advertising slogan than a land ethic.

Data supports Bunch's argument. Not only do increasingly suburban Texans have few opportunities to experience undisturbed landscapes, but a large percentage of Texans in general, and Austinites in particular, have grown up elsewhere and lack basic knowledge about its fragile ecology.

When given the opportunity, though, passionate environmentalists who actually grew up or have learned to live in the central Texas landscape describe it in terms something like the following. In the first instance, Austin was inhabited because it sits on an ecotone—the ecologically rich transition zone that mediates the flat east Texas black-loam prairie and the rugged, dry hill country to the west. This ecological division is amplified by the existence of the Balcones escarpment, a geological fault line that runs north-south roughly parallel to what is now Interstate 35. When the escarpment uplifted 300 million years ago it produced the life-enhancing springs that have supported modest settlement as well as the two hundred now-endangered species that inhabit the region. These springs are the natural outflow of the Edwards Aquifer that is recharged directly from surface water entering fissures in the limestone. "For this reason," Braune argues, "Barton Springs, the city's iconic heart, is too easily affected by development in the watershed." Today, according to former city councilor Bill Spelman, the Barton Springs section of the Edwards Aquifer provides drinking water to 45,000 people, and "local greens refer to it as the canary in the coal mine." It is in this context that Revanta too refers to the Barton Springs as "sacred"—a proxy not only for water quality in the region but for clean air, landscape quality, and the laid-back life style that has become associated with the city. Her logic is that "we all operate at a symbolic level." The environmental movement in Austin, then, is often conflated as a battle to preserve the ecological integrity of the "great springs" in general, but Barton Springs in particular.

Braune argues that the "sacred stature" of the Barton Springs "places them above science and planning." Those who see the springs in this divine light "are skeptical that we can predict and control environmental degradation. As a result, a policy of zero risk is justifiable." In this context, Braune and his colleagues reject the findings of experts—those scientists and engineers who have recommended limited development within the recharge zone. In spite of categorical rejection of expert opinion on the topic of Barton Springs, Braune recognizes that the sacred stature of the Barton Spring "has tended to obscure other urban sustainability issues, like increasing air pollution, that are equally important." A more critical assessment by Clark-Madison holds that, "instead of saving Barton Springs by saving Austin, we would save Austin by saving Barton Springs" (Clark-Madison 2002). His logic suggests that the emotional or spiritual attachment that citizens hold for the city's iconic heart tends to shield them from the emergence of harsher realities. In other words, citizens reason that so long as this one spring seems to be fine it must mean that the city as a whole is fine. Sadly, this is not the case.

Although attorney/activist Bunch clearly shares the assessment of Austin's environmental community that the Barton Springs are sacred, and in spite of his reference to Leopold, he is clearly discouraged by the long-term cultural approach to achieving sustainability that is favored by Braune and Revanta. He is, however, encouraged about achieving progress through aggressive action in the courts. Simply put, Bunch ascribes the water-quality wars to the very existence of the fragile hill country ecosystem on the one hand and the development community on the other. If the development community sees the desire of immigrants to live in the hill country as a legitimate market force that should be satisfied, Bunch sees the phenomenon as an invasion of transients made rich and ecologically imprudent by changing global economic conditions. Although it is understandable that immigrants would be drawn to the subtle beauty of this fragile landscape, he reasons that it is completely irrational to allow them to actually live there. Citing the *greatest happiness principal* of the utilitarian philosophers, he argues, "government at all levels should provide the greatest good for the greatest number through stewardship of the land." As a utilitarian in this respect—which is the political antithesis of populism—he has no qualm about suppressing the individual rights and aesthetic desires of some in favor of the rights of the majority to have potable drinking water.

Bunch's view is important because, as director of SOS for many years, he has been perceived by some as the voice of the environmental movement in Austin. Although his values may ultimately rest on a spiritual foundation, his style in attacking the development community has been hostile and expansive, not to mention expensive. He includes among the private interests that exploit nature not only developers, builders, and real estate interests but also the bank mortgage and automobile industries as a whole. Mortgage lenders, particularly in Texas, Bunch argues, encourage families to live beyond their means and privilege those forms of housing for loans that are the most environmentally damaging. The automobile industry, not to be outdone, operates within a network of road building and oil importation from the oligarchies of the Middle East that knowingly encourage suburban sprawl, water degradation, and air pollution, with all their attendant consequences. Only campaign finance reform will, in Bunch's estimation, correct the imbalance owned by the interlocking family of interests that control development in Texas and elsewhere.

In Austin, Bunch fully recognizes the claims made by the development community concerning the protection of property from illegal takings under the banner of environmental protection. But, he reminds us that, according to his interpretation of the U.S. Constitution, public rights trump private rights. As Americans, "we should not have to bribe developers to protect [our] health, safety, and welfare. Developers paint environmentalists as anti-market [actors], but they fail to acknowledge that the market is itself distorted by public investment in the infrastructure."

There is some irony in the fact that Bunch and Morford actually agree on this last point, yet their interpretation of the rationale and consequences of public investment could not be further apart. Their agreement is, as Barber would recognize, based on mutual denial that there should be any conflict at all. Both deny the validity of the other's most fundamental assumptions, or what Rorty has called one's "final vocabulary," concerning the relationship of humans and nature (Rorty 1989). The political discourse of liberal-anarchism that dominates Austin seems to breed absolute positions from which retreat is difficult.

After more than a decade fighting for environmental protection in central Texas Bunch's frustration is easy to read. Although the city "congratulates itself upon having a highly educated population," he argues, "citizens are fundamentally ignorant about the basic dynamics of urban growth." Compared to the citizens of Portland or Seattle, "Austinites" he argues "don't have a clue." He laments that "Austin has no long-term vision of its own future [no story line, because] . . . any attempt to plan comprehensively has been defeated by the joined interests of development."

Although the tactics employed by Bunch are abrasive at best, the values that drive them share a great deal with those articulated by Braune, Revanta, Best, and other environmentalists. On these grounds it seems fair to characterize the environmental counter-talk in Austin as being "green romanticism" as Dryzek (1997, 166) has defined it. And as I did for the city's dominant environmental talk earlier, green romantics can be characterized as subscribing to four related attitudes.

First, green romantics recognize the existence of global limits to quantitative growth and prefer to discuss change, if it is necessary to discuss it at all, as the qualitative development of "old Austin." Emphasis on the qualitative is consistent with equal emphasis on an inner human nature and what romantics characterize as "unnatural" practices or ideas. In extreme cases, cities themselves are considered unnatural.

Second, green romantics assume that the relationship between humans and nonhumans should be "natural," meaning that humans should swim in natural pools like the Barton Springs or Hamilton Pool rather than in chlorinated concrete swimming pools, which cause significant environmental damage. Adapting human practices to natural conditions rather than the reverse is generally articulated as an ecocentric, or nature-centered, value.

Third, green romantics argue that humans and nonhumans are both *subjects* and that neither should be exploited as *objects*. This is to say that humans should extend basic rights not only to individual animals but to ecosystems as a whole. In extreme cases romantics argue for the natural rights of any natural entity with a *telos*, or predictable developmental pattern, like that of a tree.

And fourth, those affiliated with green romantic talk employ a wide range of organic metaphors including passion and appeals to the emotions and intuition to live holistic lives. This can be accomplished, green romantics argue, by passionate defense of nature, eating only organic foods, and taking care of both spiritual and material needs without causing unintended consequences. In this context it is certainly no accident that Whole Foods Market—perhaps the nation's largest retailer of organic food products—began its life in Austin as a small grocer in 1980. In Austin's vernacular, it is this approach to life that keeps the city both "weird" and "laid back."

On the basis of the attitudes expressed by participants in Austin's environmental counter-talk, Dryzek's category seems a good fit. It would, however, be a mistake to argue that all of Austin's environmentalists hold identical views or subscribe to a single kind of talk, just as it would be a mistake to make a similar claim about the homogeneity of the development community. Both groups are diverse, held together not only by circumstance and leadership, but also by a language and conversational tensions unique to the place. Austin's environmental counter-talk—green romanticism—should, then, be understood as a rejection of Enlightenment values that is moderated by the presence of citizen experts within the community.

Austin's Technological Counter-Talk

If Austin's dominant technological discourse, *self-creation*, is adamantly technophilic, the city's technological counter-talk, *clean technology*, at least began as adamantly technophobic. By various accounts, particularly that of Austin Energy executive Larry Haupt, the origins of citizen distaste for technology lie in the antinuclear campaigns of the 1970s and 1980s. The publicly owned utility, Austin Energy, was then partly (16 percent) responsible for construction of the South Texas Project, a 1,250-megawatt nuclear facility finally constructed in 1988–1989. But, as in many other locales in the United States, construction of this nuclear generation plant did not happen without a long storm of public protest. Haupt argues that the decade-long public debate over nuclear energy was the catalyst required for the environmental movement to emerge: "That's where and when people got each other's phone numbers."

At the same time, of course, even some of the protesters against nuclear energy were engaged in building the digital technology industry in Austin—a situation that led to some very confused attitudes toward technology as a whole. There seems to have been two groups. On the one hand, some protesters—those with romantic environmental views—argued that technology is inherently dangerous because it upsets the ecological order. On the other hand, many of their comrades in the antinuclear campaign argued that digital technology is inherently good because it distributes powerful information equally without bias toward race, gender, or class.

Environmental activist Wren West has a more nuanced view. She implicitly argues that technology in general cannot be essentialized as good or bad. Even digital technology, she holds, has its environmental downside, but, where the computer chip manufacturing process can be environmentally very damaging, the material saving and creative capabilities unleashed by digital processes is entirely consistent with environmental preservation and "is a natural industry for Austin to cultivate." West's nuanced view toward technology is becoming more common, yet there remains an undercurrent of public skepticism toward technological fixes. We might say, then, that Austin's attitude toward technology is recovering from a romantic episode of technophobia.

Austin Energy executive Haupt, on the basis of his long association with the city's environmental politics, claims that attitudes toward technology have evolved significantly in the region over the past thirty years. In his assessment, the region has talked its way through four distinct conversations in that time. First was the "solar movement" initiated by the Texas Solar Energy Society (TXSES) founded in 1978. It is fair to say that TXSES was the local affiliate of the global Appropriate Technology (AT) discourse of that time. Second came talk about "alternative energy" initiated by the Texas Renewable Energy Industries Association (TREIA), founded in 1984. In looking back, Haupt recognizes that the term *alternative energy* was a particularly poor choice because it served to alienate those mainstream users who consumed the most energy but found it impossible to embrace anything "alternative." Third was talk about *sustainability* that surfaced in the late 1980s. This term is scarcely better in Haupt's assessment because no one knows what it means. The fourth kind of talk was about "clean energy." This kind of talk has his blessing (and bears his imprint).

As a student of Noam Chomsky, Haupt has become extremely sensitive to the power of language. He now understands that his job is literally to "shape the message that shapes the minds that shapes the future" of Austin. "Clean energy" and "clean technology," he argues, have become the "terms of art." If "solar," "alternative," or "sustainable" energy can be rejected by consumers as irrelevant to their everyday lives, "clean energy" cannot. Suburban values are nothing if not "clean." The evolution of language used to describe those technologies that enable us to live differently cannot, he stresses, be overemphasized.

Haupt is, then, only too aware of his role in the social construction of the city's technological counter-story line. But as a political realist he understands that the meaning of words is dependent not on what the speaker wishes for but rather the context into which his or her words are spoken. This is to say that progressive talk must be a positive interpretation of the economic and regulatory "policy drivers" of the time. To be effective, counter-talk must redescribe the dominant story line rather than invent new vocabularies of its own. Haupt, then, sees history, and the technologies that we choose, as contingent—they might have been different than what they are. Although analysts can creatively "shape language," in the end clean technologies show up in places not because great men have invented them, but because many people find a new technological language, and the artifacts shaped by it, to hold their interests (Latour 1987).

Haupt's analysis is very helpful and suggests that new language reflects broad cultural change. In this case I will argue that the appearance of "clean technology" as a coherent public conversation rests on the latent fusion of two social movements that first appeared in the late nineteenth century: the environmental movement and the public health movement. At this moment in our history a hygienic approach to technology seems perfectly natural to us, but at the end of the nineteenth century it was seemingly impossible for those who championed the interests of primeval nature and those who championed the interests of poor urban workers to see that they had anything in common. It would be fair to say that the politics of John Muir (1838–1914), the environmentalist and a progenitor of the Sierra Club, and those of Colonel George E. Waring Jr., a progenitor of public health policy in the late nineteenth century, were simply opaque and allergic to one another. What has rendered these concepts to be commensurable is the latent recognition of middle-class Americans that persistent industrialization has put at risk not only the health of distant wilderness, but their own health (Beck 1992; Moore 2005). The conscious development of new vocabularies, then, both reflects and influences social change. But

Table 2.2. The Counter-Story Line of Austin—Environmental Preservation

Kinds of counter-talk	As told by: Clark Allen, Kate Best, Glen Braune, Bill Bunch, Michelle Conn, Larry Haupt, Anna Gutierez, Sue Revanta, and Wren West
Political Counter-Talk	*Progressive Populism*
Generic response to conflict	Denial— Conflict exists only because of the certainty and inflexibility of experts
Attitude toward certainty	Skeptical— In pursuit of epistemological flexibility
What is valued	Class consciousness— From which viable communities emerge
Conceptualization of space	Jeffersonian— All yeomen deserve a quarter-section, or forty acres and a mule
Environmental Counter-Talk	*Green Romanticism*
Basic entities recognized or constructed	Global limits, inner nature, unnatural practices, and ideas
Assumptions about natural relationships	Wholeness— Relations between humans and nature should be "natural"
Agents and their motivation	Humans and nonhumans— Both groups are subjects
Key metaphors and rhetorical devices	Organicism and passion— Appeals to the emotions and intuition
Technological Counter-Talk	*Clean Technology*
Attitude toward technology	Technophobic and voluntaristic— Technology is dangerous because it upsets ecological order
Attitude toward history	Contingent— The future depends upon choices made by individuals and society as a whole
Assumptions about the origins of technology	Socially determined— Technologies emerge from public choices that can be influenced by creative language
Types of tools employed	Clean— Only tools that protect the public health

as I argued in chapter 1, such new vocabularies are less important in themselves than the story lines they serve.

Austin's technological counter-talk—clean technology—should, then, be understood as deriving from a case of technophobia but is a case moderated by a message consciously shaped to attract those citizens more concerned with public health and hygiene than in nature preservation. Austin's three forms of counter-talk are reconstructed as a whole in table 2.2.

In sum I will hold that the counter-story line of environmental preservation in Austin is constituted of three kinds of public talk: First, reformers participated in the political talk of progressive populism that rejects domination by expert culture. Second, reformers participated in the passionate public talk of green romanticism that rejects Enlightenment rationality. And third, reformers participated in the counter-talk of clean technology that emerged, at least originally, as a technophobic yet voluntaristic reaction to nuclear energy.

These observations return us to Bill Bunch's rather sobering assessment of the situation—that Austin's counter-story line, environmental preservation, has been at best marginally successful in overcoming the dominant story line of individualism. West has, I think, summed up the situation most accurately in arguing that "I would give us an A+ in opening new channels of communication [with the opposition], but a grade far lower for actually getting things done on the ground." The bitterness of the fight to preserve the environment, she holds, "has worked against us."

2.6 SUMMARY

As I argued in chapter 1, a story line is composed of several public conversations that cohere, that are logically consistent, but that compete for the attention and allegiance of citizens. Competing story lines drive history forward because they offer distinctly different views of the future. As distinct as the stories told by *individualists* and *preservationists* may seem, however, it is not likely that one group, or the story they tell, will become hegemonic. Rather, the two story lines—taken together—define a narrowed cultural horizon available for common action, if not common values. It is in this sense that the relation between the dominant and counter-conversations in a place is what I have previously described as "dialogic," by which I mean that history will unfold most satisfactorily for all parties in proportion to their ability to recognize the contested trajectory of possible story lines. This observation suggests that action is most effectively directed toward moving public conversation toward a common horizon of meaning.

In concluding this case study, it will be helpful to briefly compare the three pairs of public conversations reconstructed previously to assess how common action might become possible in the future. I will begin with assessing the two kinds of political talk, liberal-anarchism and progressive populism. Simply put, participants in both sides of this debate have been equally obstinate and grossly ideological in denying the validity of their apparent conflict. Both sides insist that they are right in absolute terms. Although progressive populists can historically be associated with healthy skepticism toward epistemological certainty, that trait has been missing at the negotiating table, at least until the arrival of Robin Rather's fresh perspective in the 1990s. As of this writing, it is conceivable that Rather and her supporters may design specific initiatives, like the Hill Country Conservancy, in which both individualists and preservationists might participate.

With regard to the two kinds of environmental talk, economic rationalism and green romanticism, there is again ample criticism for both sides. Dryzek argues that economic rationalism has been only marginally successful for at least three reasons: first, because economic rationalists fail to recognize the existence and motivations of citizens. Second, because the mechanistic metaphors employed by them are grossly reductive. And, third, because the attitude of economic rationalists toward government is thoroughly ambiguous. On the one hand, rugged individualists like Johnny Rae Morford characterize public officials as "swine at the public trough," and, on the other, Morford and his allies depend on those same public officials to maintain the conditions of contract certainty so that markets can function. Dryzek's critique of Austin's environmental counter-talk, green romanticism, is scarcely less severe. Green romantics argue that the way to change the world is by changing the consciousness of individuals, one at a time, by altering their aesthetic experience of nature. Bill Bunch's advocacy for more public access to the pristine landscape is a case in point. This is a passionate and organic worldview that seems to substitute "homo greenicus" for "homo economicus." The problem with such logic is that such microchanges do not necessarily add up to macroeffects. Put the other way around, social phenomenon are not reducible to individual psychology—society is more than the sum of its individuals. The opposition between those who rationalize the natural world as only so much stuff and those who consider it sacred seems so severe that any agreement on collective action will have to be constructed on entirely instrumental grounds such as public health.

With regard to the two kinds of technological talk, self-creation and clean technology, there is a significant opportunity that derives from our historical analysis. Clean technology talk was initially imagined in 1948, not by environmentalists but by a city council subcommittee chaired by C.

B. Smith and consisting entirely of white businessmen. The concept was based on the recommendation of their New York consultant, Richard Wood, and warmly received by the chamber of commerce, the very group that Wren West identifies as the principal obstacle to progress toward sustainability. What is so interesting about this piece of history is that, because of the intensity of their conflict, preservationists and individualists alike have been unable to recognize the common horizon they have shared for well over fifty years. As Larry Haupt no doubt understands, it has taken new language to make the recognition possible. Here, then, is an opportunity for collaborative action that might open yet others.

Glen Braune acknowledges such unlikely opportunities in his thoughtful assessment of his city. His is not a static, or final, judgment. Rather, he holds that, "although immature, Texas is an adaptable and intelligent society that sees itself as under construction." He points to the traditional friendliness of Texans to immigrants of all kinds as evidence of their future-orientation. Anyone who can contribute to the project of place building is welcome. Contrast this openness to the acknowledged closedness of New England, for example, and Texas seems a hopeful place. Braune's point is that Texans, even those who can now be categorized as rugged individualists, understand life to be developmental. If Braune is correct, this suggests that it may be possible yet to construct a Texas land ethic because Texans will see it as in their evolutionary interest to do so. Braune is not alone in his optimistic assessment. Michelle Conn made similar observations, as has Johnny Rae Morford. Morford has even gone so far as to say that he has begun to learn a great deal from Gail Vittori and Pliny Fisk III, codirectors of the Center for Maximum Potential Systems and two of Austin's most veteran and radical environmentalists. In Morford's expanded horizon, "They're great people to get ya thinkin' and I'm learning to do things different." Of course, Morford's enthusiasm in our interview might be entirely instrumental, but this does not discount the future orientation of his fellow Texans.

Finally, the line of reason employed by Braune, Conn, and Morford evokes not only the developmental theory of Charles Darwin but that of Aldo Leopold and the American pragmatists. For William James,

> [w]hat really exists is not things but things in the making . . . philosophy [he argued] should seek this kind of living understanding of reality, not follow science in vainly patching together fragments of its dead results. (James 1996)[6]

In the current context, I will interpret James's precaution to mean that we should not be too quick to finalize our assessment of how seriously the citizens of Austin take sustainability or what they are willing to do toward achieving that goal. We should, then, understand Austin the way most of its citizens do—as a city in the making—as a city still evolving. The theme of evolution toward sustainability will be taken up again in chapter 6, but even if the story line written by rugged individualists has dominated the city, and does what it can to suppress sustainable development, it has not stopped thousands of Austinites from acting sustainably. This same precaution will be useful in my analysis of Curitiba, Brazil, because it has a no less complex but very different story to tell.

CHAPTER REFERENCES

Associated Press. (2004). "Libertarian brings campaign home to Texas." *Austin American-Statesman*, November 1, p. 2.

Austin, City of. (2006). *Election History*. Office of the City Clerk. <malford.ci.austin.tx.us/election/byrecord.cfm?eid=138> (accessed 5 May 2006).

Barber, B. (1984). *Strong democracy: Participatory politics for a new age*. Berkeley: University of California Press.

Beck, U. (1992). *Risk society: Towards a new modernity*. Thousand Oaks, CA: Sage.

Clark-Madison, M. (2002). "Did SOS matter? Has the movement for clean water and democracy fulfilled its promise." In *Austin Chronicle*, August 9, pp. 22–25.

CTSIP. (2005). Central Texas sustainability indicators project. <www.centex-indicators.org/> (accessed 10 May 2005).

Davidson, C. (1990). *Race and class in Texas politics*. Princeton, NJ: Princeton University Press.

de Buys, W. E. (1985). *Enchantment and exploitation: The life and hard times of a New Mexico mountain range*. Albuquerque: University of New Mexico Press.

Dryzek, J. S. (1997). *The politics of the earth: Environmental discourses*. Oxford: Oxford University Press.

Feenberg, A. (1999). *Questioning technology*. London: Routledge.

Fischer, F. (2000). *Citizens, experts, and the environment: The politics of local knowledge*. Durham, NC: Duke University Press.

Hardin, G. (1968). "The tragedy of the commons." *Science* 162: 1243–48.

Herson, L. and J. Bolland. (1998). "Forms of municipal governance and formal participants in the urban political process." Pp. 93–129 in *The urban web: Politics, policy and theory*, 2nd ed. L. Herson and J. Bolland, eds. Chicago: Nelson-Hall.

Humphrey, D. (2003). *Handbook of Texas Online*. <www.tsha.utexas.edu/handbook/online/> (accessed 5 November 2004).

ICLEI. (2004). *Case study 5: Housing construction*. Toronto, Canada: International Council for Local Environmental Initiatives.

Imbroscio, D. L. (1997). *Reconstructing city politics: Alternative economic development and urban regimes*. Thousand Oaks, CA: Sage Publications.

James, W. (1996). *A pluralistic universe*. Lincoln: University of Nebraska Press.

Lapoujade, D. (2000). "From network to patchwork." Pp. 52–55 in J. Ockman, ed., *The pragmatist imagination*. New York: Princeton Architectural Press.

Latour, B. (1987). *Science in action*. Cambridge, MA: Harvard University Press.

Lauria, M. (1997). *Reconstructing urban regime theory: Regulating urban politics in a global economy*. Thousand Oaks, CA: Sage.
Leopold, A. (1948). *A Sand County almanac, and sketches here and there*. New York: Oxford University Press.
Logan, J. R. and H. L. Molotch. (1987). *Urban fortunes: The political economy of place*. Berkeley: University of California Press.
Menand, L. (2001). *The Metaphysical Club*. New York: Farrar, Straus, Giroux.
Moore, S. A. (2001). *Technology and place: Sustainable architecture and the blueprint farm*. Austin: University of Texas Press.
Moore, S. A. and N. Engstrom. (2005). "The social construction of 'green' building codes: Competing models by industry, government and NGOs." Pp. 51–70 in S. Guy and S. A. Moore, eds., *Sustainable architectures: Natures and cultures in Europe and North America*. London: Routhedge/Spon.
Nye, D. (2003). *America as second coming*. Cambridge, MA: MIT Press.
Oden, M. (2005). *The question of equity*. Lecture at the University of Texas, 14 April.
Orum, A. (1987). *Power, money and the people: The making of modern Austin*. Austin: Texas Monthly Press.
Osborne, J. and S. Scheibal. (2003). "The end of smart growth." *Austin American-Statesman*, 22 June 2003.
Perry, R. and K. Herman. (2000). "Perry's riding the veto record." *Austin American Statesman*, 24 August.
Portney, K. E. (2003). *Taking sustainable cities seriously: Economic development, the environment, and quality of life in American cities*. Cambridge, MA: MIT Press.
Rorty, R. (1989). *Contingency, irony, and solidarity*. New York: Cambridge University Press.
Scheibal, S. (2004). "LCRA may open new front in water wars." In *Austin American-Statesman*, 1 October, p. 1.
Sheldon, K. (2000). "Combating urban sprawl in Austin: The foundation for an effective smart growth program." Master's thesis, public affairs, University of Texas, Austin.
Spelman, W. (2002). Unpublished manuscript, author's collection, University of Texas.
Thompson, J. (2001). "Regional economy: Texas cities." <www.dallasfed.org/eyi/regional/index.html> (accessed 1 August 2004).

NOTES

1. Portney's ranking of U.S. cities with regard to sustainability was quite different than that by SustainLane (an NGO rating group) in 2005. Although the cities at the top and bottom of SustainLane's list were generally different, Austin was ranked as number 6, which is almost identical to Portney's assessment. See <www.sustainlane.com/cityindex/citypage.php?name=ranking>.

2. Howard Roark, the fictional architect created by Ayn Rand in *The Fountainhead* (New York: New American Library, 1943), p. 696. (Also cited by Barber 1984, 71.)

3. *Ejido* is the Spanish term for common lands generally used for agricultural and grazing purposes. The existence of the term documents the long tradition among Hispanic communities in North America of successfully managing part of their lands in a communal fashion. This tradition stands as evidence that contradicts Hardin's doctrine of the "tragedy of the commons." See Moore (2001) and de Buys (1985).

4. In each of the tables, I have included an accounting of those individuals from each city who have recounted their city's story line, or narrative. Narratives, however, should not be confused with people. Where narratives are internally consistent stories, people are rarely so consistent. Rather, they talk across narratives and often find themselves in strange company, or in more that one category. My purpose in listing the storytellers here is simply to help the reader recall individual story lines.

5. Friedman ran as an Independent for governor in 2006.

6. This citation by James also appeared in (Lapoujade 2000).

THREE

The Miracle of Curitiba

The city of Curitiba, Brazil, and its charismatic three-term mayor, Jaime Lerner, have achieved international recognition from multiple sources for having charted such an effective route to becoming a sustainable city: the United Nations Education, Scientific, and Cultural Organization (UNESCO); the United Nations Environmental Program (UNEP); and the International Institute for the Conservation of Energy (IICE). It is fair to say that no other city has received so much attention in the professional press for its planning initiatives. As an index of the city's public notoriety, a routine Google search in June 2005 for "Curitiba and sustainability" yielded more than 14,000 "hits" limited to English-language Web sites. This surprising number tells us something about the city's international reputation. Many readers will, then, bring with them some familiarity with the claim that Curitibanos make themselves—that theirs' is "the environmental city." (See figure 3.1.)

As I acknowledged in chapter 1, however, Curitiba's very real success story is not the by-product of democracy in the way that North Americans or Europeans understand the term. In this chapter I build on that uncomfortable observation and argue that the city's success is not the product of Lerner's singular genius but rather of a unique planning culture made possible by the concentration of power forged by a military dictatorship that held power in Brazil from 1964 until March 1985. The gradual *abertura*, or opening, of the political system after 1985 in Curitiba has been, I argue, a search for a kind of democracy that is consistent with the mode of implementation devised by urban planners with the support of military rule (Schwartz 2004).

Of course, understanding this very complex search for sustainability and democracy requires a historical background before we can attempt to analyze how Curitiba's regime of sustainability came into existence, what it achieved, and what we might learn from it.

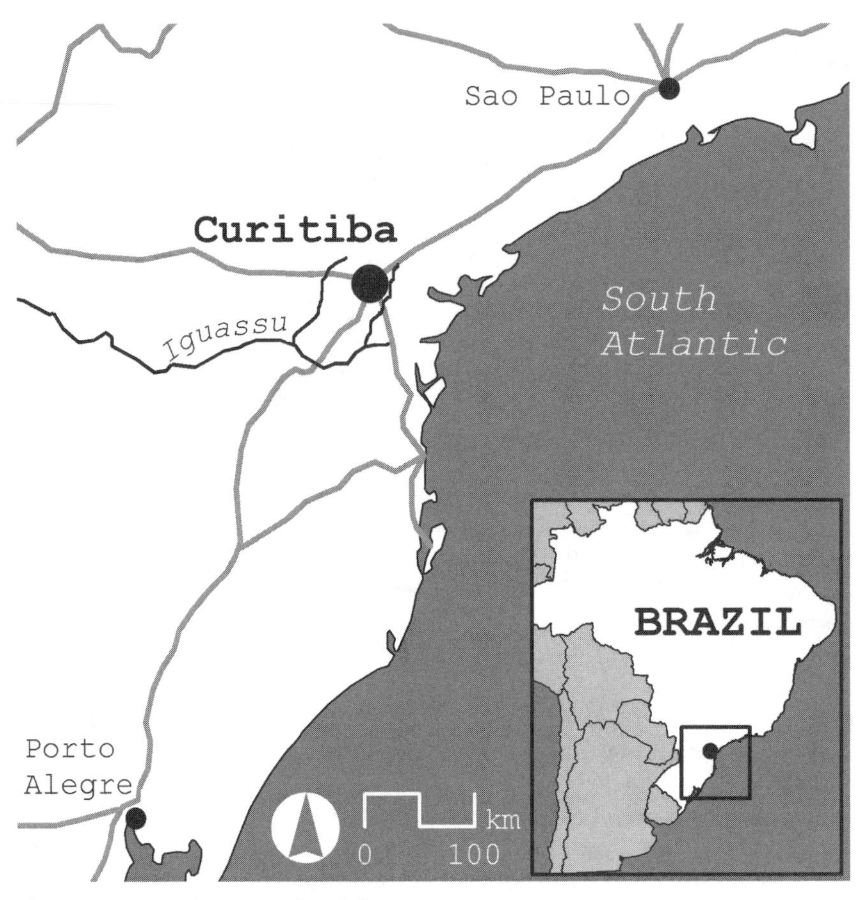

Figure 3.1. Location Map of Curitiba

3.1 SETTLEMENT AND DEVELOPMENT

The state of Parana, of which Curitiba is now the capital, was first seen by Europeans in 1530 and existed as little more than a scattering of gold-mining camps through most of the seventeenth century. It was not until 1693 when it was recognized as a distinct political region and named *Vila de Nossa Senhora da Luz e Bom Jesus dos Pinhais*, Our Lady of the Light of the Pines. That place name was, however, changed to "Curitiba" in 1721—a name that derives from the Tupi words *Coré Etuba*, or "much pine." This is a description of the countryside surrounding the city that is still accurate. The emerging city region did not achieve autonomy from the state of São Paulo until 1843 (about the time that Austin appeared on the map) but began to develop soon after. In 1978 Mayor Lerner, who is also an architect, argued that Curitiba's late development as an urban center was

Figure 3.2. The Curitiba Skyline
Courtesy of the author.

hardly atypical in Brazil and consistent with its existence within an extractive rural economy that was self-consciously antiurban. The "paternalistic" attitude toward urbanization, he argued, was first established by the Portuguese as a method to thwart anticolonial or nationalist tendencies (Lerner 1978). (See figure 3.2.)

Also like central Texas, the timing of Parana's statehood favored settlement by immigrants from Poland, Germany, and Italy because of political and economic conditions in those countries in the mid-nineteenth century. Parana's temperate climate, unlike that of Texas, was certainly another condition attractive to Europeans because of its similarity to their homelands. It was these settlers who began the transition from an economy based on mineral extraction to forestry, wood working, and agriculture, particularly the growing and processing of Paraguay tea (Xavier 1975). Curitiba is not, then, a Portuguese colonial city with a rich architecture and long history of racial mixing but a nineteenth-century "melting pot" for European immigrants.

In light of this history, it is hardly surprising that contemporary southern Brazilians describe their region as more European than the carnival-loving caricature of Brazilian society favored by North Americans. Conversely, Brazilians from the north tend to describe Curitibanos as "cold" or overly rational (Richards 1997). Ederson Zanetti, a thirty-year resident of Curitiba, describes the contemporary city as unusually conservative by Brazilian standards—a place that tends toward isolationism and does not embrace change. One should be careful, however, not to construct essential characteristics for any region in a nation as diverse as Brazil. North American readers should appreciate the kind of cultural diversity experienced by Brazilians because it is not unlike our own.

Cultural parallels are commonly drawn between Brazil and the United States but often depend on half-truths. The Brazilian constitution is, for example, explicitly modeled after that of the United States. The judicial code, however, is modeled after that of that of France. Brazil's tripartite structure of governance is, then, based on a system of checks and balances as in the United States, but the application of justice is conditioned by aristocratic assumptions. According to Roberto Kant de Lima, the hybrid nature of governance in Brazil results in a hierarchical code that tends to provide different laws for different social ranks (Kant de Lima 1995). In other words, those elites who hold public office are afforded a legal privilege unknown and unexpected by their North American counterparts.

David Hess and Roberto Damatta (1995) suggest that the social history of Brazil is rather like that of the United States south of the Mason-Dixon Line. Their analogy derives not from the bitter memory of a lost civil war, but that the social structure of power relations in both places has been more feudal than modern. As in the antebellum South in the United States, southern Brazil was the scene of an agricultural economy based on a society of landowners and field workers. But, in describing his own country, Damatta has rejected the "either/or" model of the "two Brazils" favored by foreign observers, like Tony Lloyd-Jones. In Lloyd-Jones's (1996) model, Brazilians are either very rich or very poor. Damatta, however, favors a "both/and" model. Rather than alternately rich or poor, modern or traditional, urban or rural, Damatta understands the contrasting conditions in his country as the dialectic faces of a single culture—a Janus-faced being. Armed with this dialectic model, Damatta and Hess find that, although social hierarchy tends to exist in both the United States and Brazil, the kind generally constructed in the United States north of the Mason-Dixon Line is "achieved," whereas the kind constructed in Brazil, like that in the antebellum South of the United States, is "ascribed" (Hess and Damatta 1995).

This series of distinctions suggests that, although there are unexpected similarities between the two nations, Brazilians and North Americans operate under very different expectations regarding individual rights and public responsibilities. Simply put, Brazilian elites expect privilege and citizens have very weak responsibilities. It would be a mistake, then, to interpret the activities of Curitiba's elite politicians and planners, or its citizens, through North American or European lenses.

Elite landowners in Brazil of European extraction, unlike their North American counterparts south of the Mason-Dixon Line, clung to an agrarian economy until well into the twentieth century. The industrialization of Brazil began in earnest only in 1930—very late by U.S. or European standards. The transition required a political revolution engineered by a group of semiauthoritarian nationalists, led by the *tenentes*, or younger military lieutenants, to overcome the political power of conservative landowners. The *Tenentista* doctrine was both "elitist and antipolitical" in that they believed the modernization (and defeudalization) of Brazil "could be achieved in the short run only by a totally uncompromised cadre of nonpolitical technocrats with an unswerving sense of national mission" (Skidmore 1967). The mission of modernization undertaken by the *Tenentes* was, however, better realized under the reign of the *Estado Novo* (1937–1945)—Brazil's milder version of the European authoritarian states that seized power through a military coup led by Getúlio Vargas. Unlike European fascists, the Vargas regime, at least after 1945, was more populist—more a cult of personality than an ideological regime. Thomas Skidmore has argued that Brazilian politics since the era of Vargas has been particularly receptive to such antipolitical politicians. Recent political history confirms his logic. Juscelino Kubitschek, who became president in 1955, and Janio Quadro, who became president in 1960, both triumphed on platforms that were explicitly technocratic and antipolitical (Skidmore 1967). My point in revisiting this history is that Jaime Lerner's regime of sustainability must be understood as a part of this Brazilian antipolitical, authoritarian, and technocratic tradition rather than as the political anomaly that is typically claimed by outside observers and boosters of the Lerner regime.

Even within Curitiba itself, Lerner was less the singular innovator than is commonly acknowledged by either outsiders or insiders. Father Louis Lebret, a French urbanist who arrived in Brazil in 1947 and lectured frequently in Curitiba in the 1950s, is considered by many as being the progenitor of Curitiba's innovative planning culture (Vassoler 2003). Equally significant were two previous mayors, Ney Braga and Ivo Arzua.

Braga, a graduate of the military academy, chief of police, and mayor from 1954 to 1958, went on, like Lerner years later, to become governor twice and, later, a congressman. It was Braga who arranged for the military to appoint Lerner as mayor for his first term and who appointed him for an additional term before he was ever elected by popular vote. It was also Braga who invented IPPUC (*Instituto de Pesquisa e Planejamento Urbano de Curitiba*, the Urban Planning and Research Institute of Curitiba), which proved to be the city's primary tool of planning implementation. By transforming a weak planning commission dependent on the city council into a quasi-autonomous planning agency eventually supported directly by the military, Braga conceptualized what was to become the regime of sustainability. Before appointing Lerner, Braga appointed his own wife, Franchette (an engineer), as IPPUC's director so as to better control its operation. Although most Curitibanos and outsiders credit Lerner with the invention of IPPUC, it is more accurate to hold that he was its principal beneficiary.

Mayor Arzua, elected before the military coup in 1964, was primarily responsible for recognizing the city's need for a new plan. The first comprehensive plan for Curitiba was drafted in 1943 by the French urban planner Alfred Agache. The city, however, "never had the money to implement the Agache Plan" although the plan did increase public awareness about the need to orchestrate future growth in the wake of rapid post–World War II expansion" (Cervero 1995). In 1962–1963, Arzua stimulated a year of highly democratic public talk among citizens and architects that paved the way for a well-publicized public competition to update the obvious deficiencies of the Agache plan. The competition was won by architect Jorge Wilhelm, of São Paulo, who reconceived the concentric form of the city as a series of structural axes dependent not on the automobile, as Agache would have it, but rather public transit and the pedestrian. Lloyd-Jones characterized the innovative qualities of the Wilhelm Plan by putting it in historical perspective:

> Only a few years after Costa and Niemeyer completed their pure vision of Corbusian-style functionalism in the new capital of Brasilia, Curitiba began to take the first measures aimed at reducing dependency on cars and to increase accessibility for pedestrians and cyclists. (Lloyd-Jones 1996)

Following in the wake of innovators such as Lebret, Braga, Arzua, and Wilhelm, Lerner is best understood as only one of many leaders in the movement toward innovative urban development in Curitiba. What most distinguishes Lerner within this tradition is the direct economic and political support he received from the military as well as his skill at redescribing existing conditions in a way that the conservative citizens of the city welcomed.

There is yet another Brazilian tradition that is helpful in understanding the twin phenomenon of Curitiba and the Lerner regime—that of the builder/politician. Juscelino Kubitschek began his political career as the mayor of Belo Horizonte in 1940 and quickly earned a reputation as a technocrat and builder. It was under his direction that the city became a showcase of modern architecture. When he was elected president of Brazil in 1955, Kubitschek became the conceptual, if not the design, architect of Brasilia, the nation's new capital. In fact, political careers are rather common for architects in Brazil, in part because lawyers are so discredited as corrupt (Skidmore 1967). In this sense, too, Lerner and his colleagues at IPPUC are not as exceptional as outside observers tend to claim.

It is as a part of this political history that insiders understand Jaime Lerner's success. Lerner's political career began in the political confusion of the mid-1960s, when Lerner was still a very young man and Ney Braga (Lerner's mentor) had already become governor. The electoral politics of 1964 was dominated by a military that had explicitly lost faith in the democratic process. Rather than envision themselves as the guardians of constitutional process, the military envisioned a future purged of political instability and populist mobilization of the masses in favor of a stable technocracy maintained directly by military authority. The military saw no benefit in merely guaranteeing yet another corrupt short-term regime that would inevitably lead to just another form of chaos. What was needed, in the military's view of the world, was the elimination of political and theoretical extremism coupled with competent design, capable management, and innovative solutions to real problems. The military needed "an example of urban planning efficiency for the rest of the country" (Menezes 1996). It was within this context that Lerner, already appointed director of IPPUC by Braga, was first appointed mayor of Curitiba by the military.

The Regime of Sustainable Development

Mayor Lerner was certainly not alone in his desire to solve the problems of his city. He was joined by a creative class of young designers who saw in the military assemblage of power, and the military aversion to traditional politicians, an opportunity to get things done. Architect Rodolfo Ramina characterizes the Lerner regime as having three concentric rings. At the heart of the circle is the core group that started with Lerner and who are still close to him. In the first ring are those interest groups that orbit in his gravitational pull. In the second ring are those technicians outside of government, like Ramina himself, who believe in the urban strategies proposed by Lerner and his team. What binds these groups together, Ramina argues, is their shared belief in the power of the market to transform society and skepticism about the value of public participation. On the basis of this characterization it is appropriate to associate Curitiba's regime of sustainability with the politically conservative values of neoclassical economics.

Like American journalist Bill McKibben, the young designers who orbit Lerner saw nothing wrong with an "effective paternalism"—quite to the contrary, it was the cultural norm instituted by the Portuguese (McKibben 1995). This attitude is nicely articulated by planner Sylvia Benato, who said in our interview, "The people don't know what they want, so, we do what we know is best. The people then come and celebrate."

On several occasions Lerner has argued to the effect that "cities are not as complicated as the merchants of complexity would have us believe" (Di Giulio 1994). He believes strongly that the example of Curitiba can be reproduced elsewhere, even New York City. What is needed first, he holds, is "a team of idealistic, fantastic people. Second is the simplicity of our approach" (cited in McKibben 1996, 74). Simplicity, in Lerner's lexicon, requires streamlined institutional form, what Claudino Luiz Menezes characterizes as "technological rationalism." According to Menezes, the role of IPPUC has been to rationalize policy and its implementation throughout the city. As I noted previously, this institution was transformed by Braga before Lerner's first term as mayor, but it was Lerner who seized the opportunity to expand IPPUC's authority to "coordinator of virtually all of Lerner's programs: physical-urbanistic, education, housing, and circulation" (Menezes 1996, 143). In the case of Curitiba, then, the regime of sustainability had an implementation tool distinct from city government that was economically supported by the federal military government and thus quasi-autonomous from local politics.

With institutional success, of course, also came problems. The rational simplicity that the Lerner regime brought to government in Curitiba became a beacon of light that not only attracted foreign capital investment but also displaced farmer workers from rural Parana in search of a job. The industrialization of agriculture that happened in the 1960s—the vengeance of elite landowners—put most rural citizens out of work. Lerner and his colleagues were quicker than most to understand that the old colonial divisions between urban and rural governance had come back to haunt cities in new ways.

The mass displacement of rural farm workers created new technological challenges for the city. When members of the Lerner regime such as Claudino Luiz Menezes or Rosa Moura and Clovis Ultramari describe the past forty years of development in their city, they note with a mixture of pride and veiled resentment that metropolitan Cuitiba has maintained a staggering 5 to 7 percent annual population growth rate for several decades, which adds up to nearly doubling the population every decade (Moura 1994). The concern of these political insiders is that the rate of immigration from rural Parana, as well as other regions of the country, could overwhelm the ability of planners to improvise (Filho 1997). With the statistical cards stacked against them, Curitibano planners, however, continued to rely on their pioneer spirit and faith in unorthodox, homegrown, and low-capital technological choices (Lerner 1978). Lerner, for example, argued, "I believe that the major achievement of the [municipal] authorities is the suitable selection of technology which adequately meets [the city's] needs" (Lerner 1996). Jonas Rabinovitch, a close colleague of

Lerner's, has frequently made similar claims—that Curitiba's success is based on enlightened and unpredictable technological choices (Rabinovitch 1996).

What is interesting about these characterizations of Curitiba's technological success is not the emphasis on technological artifacts per se but on those responsible for innovation. In these examples and others, credit is invariably given to what Lerner characterizes as a small "team of planners, technicians, and architects" (Di Guillio 1994, 84). This is to say that Lerner and his colleagues fully understand that technologies not only are objects, but may include systems of people, knowledge, and social practices—a "national culture" (Lerner 1978, 13). In their view it takes an expert to make important choices on behalf of less-informed, sometimes illiterate, citizens. And it is this cadre of designer-experts on whom disengaged citizens depend to manage the complexity of urban life and growth.

My point in this historical analysis of IPPUC and the Lerner regime is that the democratization of Brazil at the national level after 1985 has not rid Curitiba of the "clientelistic system" that dominated politics in the era of the *tenentes* of the 1930s and before (Vassoler 2003). Contemporary citizens, who see themselves as having little responsibility to improve the world, depend on technocratic elites to do the job on their behalf. In this context we might still think of Curitibanos more as "clients" of the Lerner regime of sustainability than as "citizens" of the city.

3.2 THE CLIENTS' VIEW

At the opening of this chapter I argued that no other city in the world has received so much global attention because of its public aspiration to develop sustainably. The American economist Hugh Schwartz, for example, has characterized Curitiba as "what may be the most successful urban renewal of any sizable municipality in the world during the late twentieth century" (Schwartz 2004). Such praise from distinguished outside observers is extremely important because it reflects two things: an expert opinion that, first, the city is a good place to invest and, second, the city's administrative apparatus may be a model that could be exported elsewhere. In both contexts it suggests that Lerner and his colleagues understood how foreign observers, as well as the city's inhabitants, were "clients." It will be helpful, then, to review what both external and internal clients have found so impressive.

Schwartz is not alone in his use of hyperbole. Foreign observers such as Tony Lloyd-Jones (1996), Robert Cervero (1995), and Julian Hunt (Hunt 1994) give Curitiba credit for creating Brazil's highest uniform standard of living and lowest unemployment rate. This achievement seemed extraordinary to these observers because, by comparison, most other Brazilian cities suffer from the simultaneous extremes of gilded wealth and grinding poverty. Hunt (1994, 2) goes so far as to claim that Curitiba is a "new space" in the context of Brazil that deserves wide attention. Richard Rogers, the well-regarded English architect, agrees. In his 1995 Reith lectures, Rogers characterized Curitiba as an example of "sustainable urbanism" miraculously transformed from "a collection of *favelas*, or shanty towns," into a model for cities around the world to emulate (Lloyd-Jones 1996).

Paul Hawken, Amory Lovins, and Hunter Lovins also employ the city as an exemplar of "comprehensive planning" (Hawken et al. 1999). In their description of Curitiba's many successes, they characterize it as benefiting from both "responsible government and vital entrepreneurship" (Hawken 1999, 288). The authors further argue that the city's innovative brand of government and entrepreneurship have lead to "measurably better levels of . . . democratic participation" (Hawken et al. 1999, 382). This generally positive if unfounded characterization of the city's efforts is, perhaps, best typified by Bill McKibben.

McKibben's *Hope, Human and Wild* documents three examples that might serve as heroic, but nonutopian, models of sustainable development—the forests of the northeastern United States; Kerala, India; and Curitiba, Brazil. In his investigation, McKibben "decided, with great delight, that Curitiba is among the world's great cities" (1995, 61). In his assessment, that distinguished status is achieved not because of great architecture, remarkable landscape, or exceptional food but because of the style of its politics. During his month-long stay in the city, McKibben claimed, "I met very few cynics. The resigned weariness of Westerners about government, which leaves only fanatics and hustlers running for office, had lifted from this place." In his chronicle, the enthusiasm for city building that he found among his hosts "helps redeem the idea of politics" (McKibben 1995, 62).

Outsiders like McKibben have found much in the Brazilian political economy to be celebrated. In mid-1994 Brazil adopted the *Real* Plan (the real is Brazil's unit of currency) that finally stabilized the nation's chronic inflation problem. Economists and historians have commonly characterized twentieth-century Brazilian economic policy as undisciplined or even irrational (Skidmore 1997). Although that judgment may have merit for the nation as a whole, the reverse appears to be the norm in Curitiba. Schwartz (2004) points to the disparity between São Paulo, 210 miles to the northeast, and Curitiba as one reason for the latter city's success. Whereas São Paulo, the nation's largest city, has been plagued by pollution, crime, and congestion, Curitiba has earned a reputation for orderliness and discipline. As a result, many corporations have abandoned the chaos of São Paulo to feed that market from the more secure conditions of neighboring Curitiba. In this sense these outside supporters are also clients of the regime who support its continued operation.

In the late 1990s, about 35 percent of Curitiba's workforce was employed in the retail, commercial, and service sectors and 19 percent worked in the manufacturing sector. The latter statistic, however, was given a significant boost at the end of the century by the location of major manufacturing operations by Fiat, Pepsi, and Volvo into the city (Cervero 1995). This kind of rapid economic development has been associated by many with the dramatically dropping birth rate—certainly an important sustainability indicator (Filho 1997).

As noted earlier by Brazilian observers, foreigners, too, recognize there is a price to be paid for so much success within a generally less robust economy. Eugene Linden characterizes Curitiba's situation regionally:

> Ironically, the very programs that have made Lerner one of the most popular mayors in Brazilian history threaten Curitiba's future. Says Ashok Khosla, president of the New Delhi–based Society for Development Alternatives: "Each city contains the seeds of its own destruction because the more attractive it becomes, the more it will attract overwhelming numbers of immigrants." Luciano Pizzato, a federal deputy from Curitiba, notes that during the next ten years, Brazil's population will grow by 40 million people—an increase the size of Argentina's population. "You cannot create facilities for a new Argentina in ten years," says Pizzato, who fears that Brazil's poor will make Curitiba their destination of choice. (Linden 1996)

The industrialization of agriculture in southern Brazil during the latter half of the twentieth century is directly linked to urban immigration and accounts for a population increase in Curitiba from 500,000 in the mid-1960s to 2.4 million in the late 1990s, with 2.7 million forecast for the year 2020, making Curitiba the seventh-largest city in Brazil. Even the enthusiastic McKibben (1995) noted that displaced farm workers from the surrounding state of Parana, as well as other regions of the country, have inhabited some 209 *favelas* that constitute more than 10 percent of the city's population. Susan Di Guilio (1994), writing for the American professional journal *Progressive Architecture*, saw immigration as the biggest threat to the city's "sustainability."

Other observers—the transportation planner Robert Cervero (1995), for example—seem more convinced that the strong planning tradition in Curitiba could survive the onslaught. Population growth had, after all, been constant throughout the city's long planning process, and seems to only have stimulated creativity. The dominant view of Curitiba, like Cervero's, rests on faith in the planning culture that has been developing in the city for fifty years.

Although the transit-based nature of the form conceived by architect Jorge Wilhelm, winner of the 1966 competition, has received the most attention outside Brazil, in Cervero's view

> [i]t is important to note that in Curitiba, larger land-use objectives drove transportation decisions, and not vice versa. Planners and civic leaders first reached agreement on what physical form the city would take—a linear one that would achieve more balanced growth and preserve the social and cultural heritage of the central city. Realizing the "derived" nature of travel demand, transportation decisions were then made to reinforce land-use objectives—namely, building axial transitways that would help guide and serve the linear growth. (Cervero 1995)

Cervero's point, then, emphasizes that the urban planning process employed in Curitiba was unique. A similar argument is constructed by Julian Hunt (1994, 3) in the popular design magazine *Metropolis*. He holds that where other Brazilian cities allowed intensive development to dictate urban form

> Curitiba managed something that no others did. . . . The city effectively carried out a plan that included transit, housing, parks, and public spaces organized in a comprehensible urban shape.

Thus, on the basis of reports by North American and European observers, Curitiba seems so miraculous because its approach to planning appears to be comprehensive—an approach that no other North American or South American city has realized.

Of course, these testimonials by outsiders require some explanation of why Curitiba seems to have succeeded in planning comprehensively where other Brazilian cities, subject to similar geopolitical conditions, have failed. McKibben (1995), like Schwartz (2004), credits the architectural culture of the city as having developed a unique ability to recognize and solve problems. I will argue, however, that this homegrown planning culture is best characterized not as "comprehensive" in the all-inclusive or positivistic sense, but as "ad hoc." Rather than broad, methodical, and linear, planning in Curitiba has been conducted as a series of small campaigns of limited scope, not a single integrated or "comprehensive" movement. Jonas Rabinovitch (1996), director of international relations for the city in the 1990s has, for example, described his city's planning process as an "action script" as opposed to the kind of rigorous quantitative policy analysis generally practiced by economists. This characterization is reinforced by Schwartz (2004), who was surprised by at least three aspects of the city's way of planning: First, he found that economic and physical planning was conducted by informal, or "rule-of-thumb," methods. Second, he discovered that few people were involved in both determining and implementing policy. Schwartz's third surprise was learning that this highly centralized, ad hoc method was so successful. In comparison to North American and European norms, it seems to outsiders that the architect-politicians of Curitiba have achieved a miracle on behalf of citizens.

McKibben does recognize that many of Curitiba's "action scripts," like *Tudo Limpo*, or "all clean," are "paternal" because they are designed by technocrats to alter social behavior. He argues, however, "what makes it seem all right is that it's an *effective* paternalism, not a sham" (McKibben 1995, 86; emphasis in original). In McKibben's assessment, we should be willing to overlook the use of what appear to North American and European readers as undemocratic means because the regime of sustainability is actually getting the job done.

Five Innovative Strategies

Getting the job done has been made possible by implementing five innovative strategies: the production of green space; creating the "structural axes" mentioned earlier; the much-heralded transit system; attracting foreign investment; and what I call the city's *incremental projects*. It is these strategies that have so impressed foreign, as well as local, clients, and I will briefly visit each of them in turn. In fairness, however, I will also document briefly the existence of counter-interpretations that will prepare the reader for the next section, "The Citizens' View."

The first strategy adopted by the Lerner regime was solving multiple environmental, social, and economic problems through the production of green space. When Jaime Lerner first took office in 1971 there was only one park downtown, the *Passeio Público*. By 1992 "Curitiba boasted fifty square meters of green space per citizen—four times the World Health Organization's recommended minimum" (Boles 1992). In chapter 5 this claim comes under scrutiny, but this seemingly remarkable transformation was achieved not simply to improve recreational opportunities for citizens in an increasingly dense city but also to alleviate property damage caused by chronic flooding of the Iguaçu River. Through an aggressive program of property condemnation, rezoning, and public works projects, IPPUC reports that the floodplain was transformed from a health and safety liability into an urban asset (Hunt 1994; Richards 1997).

There is, however, another view. Paulo Pereira—an architect employed by a Curitiba-based environmental nongovernmental organization—recognizes some truth in IPPUC's claim. Emphasizing the number of square meters of park space available to each citizen, however, tends to suppress the data that document the unintended consequences of planning. In the case of the floodplain parks, the landless workers displaced from this land have now moved just outside the city's boundary to land that is even more environmentally fragile and thus threatened by unplanned use. In Pereira's view, the reality of concern to citizens is suppressed by statistics.

The second strategy employed by planners in Curitiba was to implement the structural axes envisioned by the Wilhelm Plan. These were constructed during the 1970s. The linear forms of the axes constitute what planners call a "trinary system" in which a major street served by high-speed and high-capacity buses is flanked by two secondary streets that form an automotive "couple," or alternating one-way traffic arterials. Zoning designates that density decreases as one moves away from the center of the axis. Thus, between 1970 and 1992, average gross residential densities along the axes increased 855 percent and approached 100 dwelling units per hectare, or 38 dwelling units per acre. Close to downtown, densities of 170 units per hectare were achieved (Cervero 1995, 17). The clarity of the city's urban form is a secondary benefit of the structural axes that drives it. Such a linear or starburst form is, of course, not new in itself. It is clearly derived from a theme that historians associate with Soria y Mata's 1894 proposal for Madrid. What may be new, however, are the densities achieved in Curitiba. Tony Lloyd-Jones (1996) argues that although such a high-density starburst form graphically implies political centralization, the Lerner regime has attempted decentralization by creating nine secondary mixed-use centers along the axes where development is encouraged by the construction of "local, decentralized town halls, sports and cultural facilities, as well as commercial service." These multifunctional centers are supported by the proposal to construct some fifty new "lighthouses"—symbolically patterned after the ancient lighthouse of Alexandria—that contain libraries and Internet connections. It is fair to say that the urban form constructed in Curitiba appeals to outside observers not only because it is clear, decentralized, and efficient but also because planners have been able to integrate public services and transportation with morphology in a remarkably short time. (See figures 3.3 and 3.4.)

Again, there is another interpretation by those interested in the greater good. Curitiba architect Rodolpho Ramina agrees that the densities achieved are incredibly efficient, but he also holds that such densities have become both unaffordable and uninhabitable. Although a supporter of the Lerner regime in theory, he has decamped his own home to the suburbs.

The third strategy employed by Curitiba—its Integrated Transportation Network (RIT)—has been managed by a public-private company, Urbanização de Curitiba (URBS) since 1974. As early as 1994, Hunt documented that the system served over one million riders per day at a cost of three hundred times less than a comparable subway system system (Worcman

Figure 3.3. An Example of Curitiba's "Structural Axes"
Courtesy of the author.

1993). Although planners in Curitiba began to consider the construction of a subway or monorail system in 2000, they argue that the city could not have afforded such an expensive infrastructure investment thirty years ago when residential densities were dramatically lower. In that period the bus system "captured 28 percent of the city's car owners daily" (Di Giulio 1994) and satisfied 70 to 80 percent of the daily trips taken by Curitibanos (Boles 1992, 59). This high degree of ridership resulted in fuel consumption rates 25 percent lower than other Brazilian cities of its size accompanied by comparable pollution and respiratory disease rates—certainly another indicator of sustainable development (Lloyd-Jones 1996). The popularity of the bus system reflects the ease of its use afforded by clear signage, frequent connectivity, low cost, and reliability.

The citizens' view, however, has compiled credible evidence that Curitiba's increasing economic dependence on the automobile industry and the informal partnership of the RIT board of directors and real estate interests have greatly compromised the transit system's ridership and effectiveness (Irazabál 2005).

The fourth strategy to be designed by Curitiba's planners was development of the CIC, or the Cidade Industrial de Curitiba. By condemning forty square kilometers of property in the southwestern part of the city and placing its development under strict government control, Lerner and his colleagues were able to complete the infrastructure for the CIC in a

Figure 3.4. Lighthouse of Knowledge
Courtesy of the author.

two-year period ending in March 1975. By providing very attractive financing, the regime was eventually able to attract some 415 companies to locate there. According to Lerner and his supporters, the strict environmental regulation imposed on these investors was an attraction rather than a liability, because regulation contributed to a climate of international contract certainty (Katz 1997).

Schwartz (2004) argues that the CIC initiative was highly successful for five reasons: first, because of public relations; second, the land was cheap; third, the plan included expansion space; fourth, government financing and infrastructure assistance was made available to international corporations; and fifth, local labor costs were lower than in São Paulo, the closest industrial center. By linking the CIC to the transit network, the local labor force was made instantly available.

But here, too, citizens report that newer industries—Renault, in particular—have shunned the CIC location in favor of less expensive land in environmentally fragile areas. Remarkably, such deals have been made with the full support of the Lerner regime (Irazabál 2005).

The fifth strategy designed by Curitiba's planners is what I will call its *incremental projects*. These are the small-scale, quick-return campaigns stimulated more by imagination than by the deep financial resources required to plan and implement basic infrastructure. Many observers have argued that the period of heroic infrastructure creation was over in the

Figure 3.5. A Rapid-Transit "Tube Station"
Courtesy of the author.

city prior to 2000 and that it is these socially inspired projects that will make the most important contribution. Although the incremental projects are too numerous to provide a comprehensive assessment within the scope of this book, the mention of a few will help to characterize the general approach.

Perhaps the most famous of the incremental projects is the pedestrianization of Rua das Flores (Flower Street), now a major shopping district in the central city (see figure 3.6). This was a project at least implied by the Wilhelm Plan but executed by the Lerner regime in a manner that seems remarkable to outsiders. Over the strenuous objections of local merchants, and with the help of armed police, Lerner barricaded, repaved, and planted the street in a single weekend before merchants could mobilize an effective protest. In response to the merchants' preempted protest, Lerner took the impressive risk of promising that if the merchants were not happy in six months, the government would restore the street to its former condition (Schwartz 2004). When the new pedestrian precinct was quickly overwhelmed with beggars, Lerner countered not by passing vagrancy laws enforced by the police but by employing the beggars to keep the "mall" clean. Robert Mang has characterized this project as an example of the "value-added, reciprocal relationships" that make Curitiba work (Mang 1995). According to official records made available by IPPUC, the

Figure 3.6. ***Rua das Flores*** **(Flower Street)**
Courtesy of Jorge Mata Otero.

district is now a flourishing urban pedestrian mall welcomed by the same merchants who originally opposed it. Some local observers, however, still characterize this method of implementation as a brilliant but heavy-handed method of conflict suppression (Ramina 2005).

Daralice Boles, writing in *Landscape Architecture*, has written enthusiastically about the city's "green exchange" as an exemplar incremental project (Boles 1992). Because the *favelas* that house most of the city's poorest citizens cannot be serviced by large modern garbage trucks, planners conceived a way to get residents to bring the garbage out, thus improving sanitation and reducing environmental threats to health. In exchange for a given volume of garbage, citizens are compensated with food—which is necessary to live—or bus tokens—which are necessary to get to work. In this green exchange, both farmers and poor citizens benefit at very little cost to the city. A not insignificant benefit reported by IPPUC is that 70 percent of Curitibano families recycle their garbage compared to 15 percent in suburban Toronto, which is typical in North America (Boles 1995). But again, locals like architect/planner Rodolpho Ramina provide contrary evidence about the efficiency and use rate of the entire recycling program. In his wry view, the program was developed "for English eyes" only.

Incremental projects also sprang up in the realm of health care. Curitiba, before most other Brazilian cities, implemented a federal law that prescribes the decentralization of medical care. In 1996 there were eighty-six outpatient clinics in the city open twenty-four hours a day that provided rudimentary care and referrals to thirty-six full-service hospitals (Cavacanti 1995). For those orphaned street children who are present in every Brazilian city, Curitiba implemented an adoption plan that associated children with a business. In return for a daily meal and a small wage, children help provide routine maintenance, care for gardens, or do office chores (Meadows 1995).

To combat mass immigration, the city established two related incremental projects: the Projeto Vilas Rurais (Rural Village Program) and the Projeto Bom Negócio (the bus ticket plan is part of the Bom Negócio project). Begun in 1994, the return house project provides free bus tickets to rural immigrants to return to their village. In 1996, with the aid of the World Bank, the state of Parana constructed a series of rural villages intended to retain peasant families in the countryside, thus avoiding some urban immigration. When originated in 1996, fifteen villages were under construction and sixty were being planned with a target population of 50,000. Landless peasants were given a 5,000-square-meter agricultural plot and a simple house in exchange for the pledge of long-term residence—a kind of "homesteading" agreement that will be familiar to American readers. In the eyes of the landless peasants themselves, as reported through the *Movimento dos Trabalhadores Rurais Sem Terra* (Landless Workers Movement, MST), this project proved unsuccessful because it proved to be yet another case of conflict suppression. By keeping the rural poor out of Curitiba, and out of the *favelas* that breed dissent, they could be better and less expensively controlled.

McKibben reported on a number of incremental projects including the "cohab" infill housing project and the Novo Bairro subdivision in which settlers constructed their own houses. To help settlers with the design and construction process they were given a one-hour consultation with an architect and access to a "technology street" where various sustainable construction techniques and innovative designs were on display (McKibben 1995). The emphases on self-help and environmentally sound construction are messages that were made available to the general public as well as to state and municipal employees through educational programs offered by the Free University of the Environment established by Lerner in 1991. Even this landmark of the regime, according to Ramina, is deeply flawed. Although the structure is reported to be an example of sustainable architecture, closer scrutiny reveals that its primary environmental asset is that it was constructed of eucalyptus logs—a romantic allusion to the era of settlement and hardly an efficient or sustainable use of material. (See figure 3.7.)

In spite of local critiques by citizens, each of the five strategies employed by the Lerner regime—the production of green space, the structural axes, the transit system, attracting foreign industry to the Cidade Industrial de Curitiba (CIC), and the incremental projects—enjoyed considerable local, national, and international acclaim as being innovative and sustainable.

Whole-Systems Thinking

The innovative nature of these five development strategies had a great deal to do with the expanding reputation of the Lerner regime. So much coordinated innovation, of course, requires political power. Many reporters from the outside, McKibben in particular, assigned singular genius and power to architect/mayor Jaime Lerner. Other more sophisticated observers, such as Donella Meadows (1995) and Hawken et al. (1996), argue that Curitiba's success can be attributed to not one but a continuum of six designer/mayors who have treated "transportation and land-use, hydrology and poverty, flows of nutrients and of wastes, health and education, jobs and income, culture and politics, as intertwined parts *of a single integrated design problem*" (Hawken et al. 1999, 293; emphasis in original). If one analyzes this claim in any detail, it suggests that "design," as proposed by Hawken, Lovins, and Lovins, is not the graphic description of a physical artifact but the purposeful integration of human and natural systems. It is the design of a way of life.

Outside observers like Hawken et al. admire the political continuity constructed in Curitiba because it permits what they and others refer to as "whole-systems thinking" in a way that is impossible to imagine either in North America or in other Brazilian cities. Whole-systems thinking is impossible in the United States because of the individual property rights restrictions that we saw in Austin and because control over various urban infrastructural systems is dispersed between competing agencies. But in most other Brazilian cities—except Porto Alegre—such integrated planning has proved practically impossible, not because of property rights issues or the atomization of power but because of the wild policy shifts endemic to four-year electoral cycles (Vassoler 2003). In Brazil there are

Figure 3.7. The Free University of the Environment

Courtesy of Jorge Mata Otero.

many political parties with a wide spectrum of ideological and personal affiliations and it is common for cities to start projects only to have them abandoned by the next regime. Hunt (1994, 6) is even more direct, arguing that the political stability that has enabled the success of integrated planning in the city derives directly from the prior military dictatorship. IPPUC, the Urban Planning and Research Institute of Curitiba, has "maintained the continuity of the urban projects" in a way that "the city cannot function without it." Christina Cavacanti (1995) makes a nearly identical argument that credits an efficient technocracy with creating the political continuity required to foster sustainable development. From the remote perspective of these observers, then, whole-systems thinking depends on a single authority having control of *all* systems.

Hawken, Lovins, and Lovins (1999, 20) admire not only the efficiency of Curitiba's planning culture but also what they describe as their "non-ideological approach to problem solving." For these authors, the Curitibano designer/politicians have constructed a "society [that] can work toward resource productivity . . . without waiting to resolve disputes about policy." Toward the end of their book Hawken et al. (1999, 321) go so far as to conclude that "to make people better off requires no new theories."

Having begun this book by arguing that technology is inherently political, it is hard to accept the reasoning of Hawken and colleagues. The irony in their argument is that they fail to appreciate that their objection to theorizing and policy debate is itself a theory. It is one thing to overlook ideological differences in the interest of getting things done, but it is quite another to condemn theorizing and debate altogether as useless and obstructive activities (Light 1996; Rorty 1998). We will return to this issue in chapter 6. For the moment, my point is that the advocates of "whole-systems thinking" cannot have it both ways. They cannot argue on the one hand that design is an expanded activity that integrates human and non-human systems as a way of life and then argue on the other that design is apolitical. Their position is either confused or is itself an attempt to suppress conflict—an activity sure to incite dissent.

Before reconstructing the counter-view of dissident citizens, I must make clear there are many locals who see the regime of sustainability as acting in their interest. If this were not so Lerner would not have been elected to his third term, nor would Mayors Greca, Taniguchi, or Richa (all associated with the Lerner regime) have been popularly elected in Lerner's wake. The assessment of the Lerner regime by those who see themselves as citizens is, however, more complicated than that of those who see themselves as clients because they bring with them an engagement with the society as a whole that clients lack.

Missing Evidence

The overwhelmingly positive assessment of Curitiba's development by the regime's supporters from both the outside and inside is indeed founded on evidence. Before considering how more democratically inclined Brazilian observers interpret the miracle of Curitiba, however, it is necessary to at least comment on evidence that is missing from my review of the literature related to the city.

Although some outside observers (Hawken et al. 1999, in particular) have explicitly argued that the city is an exemplar of democratic inclusion, they present no evidence to substantiate this claim. Nor do any outsider clients adequately explain how Lerner and his colleagues concentrated the political power required to plan in such an integrated manner. How, for example, did IPPUC achieve the radical rezoning of the entire city in 1972 and again in 1975 that enabled production of the city's coherent form? Nor is there any consideration of the political process utilized by the municipality to condemn and acquire the private property that became the CIC, or Cidade Industrial de Curitiba. The planners of Austin would be nothing less than astonished by the authority assembled by their counterparts in Curitiba. Nor do these enthusiastic authors document how the city administration employed the police to manage and control projects like the redevelopment of Rua das Flores (Flower Street) and the construction of the rural villages. These are all development techniques that are unthinkable in North America or the European Union, but are admired for their "efficiency" by some North American and European supporters in the context of Brazil.

In summation, I will argue two points. First, the regime of sustainability in Curitiba exists as a "clientelist system" as did the various military regimes of Brazil's past (Vassoler 2001). As I argued earlier, this characterization suggests that business leaders, politicians, and planners form the elite class who are granted authority to govern by their internal and external "clients" so long as "the trains run on time." What is unique about the Lerner regime of sustainability is that it has been successful for more than forty years in keeping the urban infrastructure running efficiently. This is not to say that it is entirely undemocratic because the system takes care of and listens to its own. In this limited sense it is democratic.

Second, I will argue that the Lerner regime has been extraordinarily successful at telling stories to both its internal and external clients. One genius of the regime has been to provide foreign observers with information so attractive that they would reproduce the city's story in the banking centers of North America and Europe. This is to say that Lerner and his colleagues managed to redescribe a rather conservative and provincial town to be the urban model for the future—the "ecological capital of the world." This is a remarkable accomplishment.

Data, Propaganda, and Stories

Precisely because the regime of sustainability was so successful in its redescription of Curitiba, it is necessary to make a distinction between *data*, *propaganda*, and *storytelling*. Most readers will, I think, recognize that the line between these categories is often blurry. We have become sophisticated consumers of statistics and generally recognize that numbers are capable of telling lies because they quantify reductive variables and erase information that may be significant to citizens yet resistant to quantification. The most sophisticated observers recognize that quantitative data reflects little more than the assumptions of a particular community of inquirers and the theoretical presuppositions to which they subscribe (Fischer 2003). On the basis of this logic, I will hold that *data* document only a selected portion of the past—the part we can quantify. This is the kind of reality constructed by scientists. In contrast, *propaganda* is the purposeful distortion of data—the projection of a knowingly false reality for political purposes. This is the kind of reality constructed by corrupt politicians. And finally, a *story line* is the responsible speculation of what might become true in the future—it is a hope for a better future, but one founded on history. This is the kind of reality proposed by prophets, for instance, individuals such as the Reverend Martin Luther King Jr.

If we accept this generally postmodern attitude toward reality and quantitative data, it is necessary to ask at what point does our enthusiasm to tell compelling stories—those that will help to bring about a brighter future—become dishonest political manipulation? This is an ethical question that I will not attempt to answer here. I will, however, argue that all data is to some degree propaganda and that most propaganda has something hopeful, even if misguided about it. In the case of Curitiba's regime of sustainability there is ample documentation that some of the data generated by Lerner and his colleagues were certainly closer to propaganda than hopeful story lines. The most sophisticated Brazilian critics of the Lerner regime, such as social scientists Márcio de Oliveira and Pedro Jacobi, are highly skeptical of the self-serving data produced by IPPUC and by Lerner protégé Jonas Rabinovitch. De Oliveira, in particular, has reconstructed the rhetoric of sustainable development employed by the Lerner regime and found it to be a masterful, if inaccurate, redescription of history. But, even such enthusiastic outside observers as Bill McKibben (1995, 58) recognize that some of the city's programs "are so over-rated as to be classified as 'propaganda.'" As in any political story, of course, there is a host of detractors who have accused the Lerner regime of fraud (Samek 2001), brutality (MST 2004), bad faith (Schwartz 2004), and a host of other crimes (Editorial 1997a; Editorial 1997b; Segwaw 1998). Merely repeating those accusations in detail without adding substance to them, however, is not particularly helpful. Ederson Zanetti, for example, brushes away such accusations by saying these problems are not unique to the Lerner regime but are a general cultural problem related to the values of weak citizenship.

One claim made by citizens to which this book can productively contribute, and which is central to the questions asked in chapter 1, is that the Lerner regime has practiced a particularly opaque and secretive form of governance. Lerner himself has argued that some secrecy in planning the city was required to avoid real estate speculation in areas being redeveloped (Lerner 1978). Some detractors, however, argue that secrecy only contributed to corruption and that the need for secrecy was a thinly veiled excuse to suppress dissent (de Oliveira 2001). These written accusations are supported by interviews with locals. In any case, there seems to be little disagreement on the details—the Lerner regime got things done behind closed doors in a way that most American or European institutions would never accept—even in the name of efficiency.

3.3 THE CITIZENS' VIEW

If it is not already implicitly clear, I should further distinguish between clients and citizens in a democratic state. *Clients* are those residents of a place who are dependent on protection and services provided by municipal experts, who are in turn compensated by votes on which they depend to stay in power—the relationship can be characterized as the exchange of favors between individuals. As I have argued previously, however, regimes can also have nonresident clients who similarly benefit from the exchange of favors. In contrast, *citizens* are those residents of a place who are granted rights and responsibilities by government in exchange for loyalty to the greater good. The distinction is twofold. First, clients can never be experts in the political realm, but citizens are always both givers and receivers of expert service and protection. Second, clients exist in a world maintained by individual bargains and citizens exist in a world where the collective good is considered greater than the sum of individual goods. If we accept these distinctions, it means that clients and citizens will interpret reality quite differently.

An example of this distinction is that the popular press reviews of Curitiba's remarkable planning history by clients of the regime never mention the existence in the city of a counter-story line. I have already mentioned that Hawken et al. (1999) go so far as to discredit any and all competing interpretations of reality as "ideological" and thus an unhelpful return to Cold War–era partisanship between the Left and Right. These claims by outsiders, of course, do not reflect the reality seen by those Curitibanos who fought very hard against the military government to achieve citizenship in the sense intended here. Their perspective is far more nuanced because it is historical and developmental.

Periods of Planning in Curitiba

A number of architects and policy makers in Curitiba (particularly Rodolfo Ramina, Paulo Pereira, and Roberto Allende) have contributed to my reconstruction of Curitiba's planning history from the public record. Of course, not all observers agreed on various details and time spans, but overall I found substantial agreement on content that differs a great deal from the monolithic history articulated by regime clients. These observers, citizens all, articulate four periods of modern planning that roughly correlate to the decade following the military coup: small is beautiful (1965–1975); modernization (1975–1985); redescription (1985–1995); and metropolitan reality (1995–2005). Brief characterizations of each period are helpful, but together they establish a developmental trajectory that is more important that any individual period itself.

The first period (1965–1975), when small was beautiful, is by now something of a heroic myth of origin when, as Roberto Allende puts it, "implementation was faster than ideas." The industrialization of agriculture (supported by the military on behalf of landowners) created mass worker displacement and a corresponding urban immigration problem that could be solved by creativity disciplined by a rather simple formula: (transit + zoning) = (density + efficiency). The required discipline came from streamlined methods of implementation created through IPPUC. This formula worked better than almost anyone expected.

The second period (1975–1985) was an era of infrastructure modernization "when ideas and implementation traveled at the same pace." Bigger projects, of course, take more time—especially during a national economic crisis precipitated by the international monetary policy followed by the military. This period ended with the beginning of national redemocratization, a new constitution, and increasing awareness of the global markets in which Brazil would have to compete. At first, the phenomenon of globalization seemed an opportunity to level grossly inequitable income distribution within Brazil and around the world. That initial promise, however, soon faded when income concentration began to increase rather than decrease (Suplicy 2003).

The third period (1985–1995) was marked by redescription—of making and marketing the city's identity. In lieu of building infrastructure still needed by many citizens, IPPUC redecorated the city with ethnic memorials and fanciful architecture that celebrated the city's ethnic diversity, European ties, and commitment to "greening." In this period ideas traveled faster than implementation because, as Ramina puts it, "greening was taken almost literally by IPPUC" rather than substantively. He means that architecture and landscape were employed for their representational value rather than in a systemic way that would actually reduce consumption. In other words, planting a few trees was emphasized over reducing massive carbon emissions from South America's largest cement plant. And because the new constitution adopted at the beginning of this period gave a great deal of power to cities, it became extremely difficult to coordinate planning between municipalities in the region. As a result, the rising cost of living within the city forced very large numbers of people into *favelas* just outside the city borders.

The fourth period (1995–2005) was, according to Allende, one of metropolitan reality—"when beauty and efficiency can no longer be small." It was in this decade, with Lerner as governor, that it was again possible—as it had been in the military era—to coordinate city and regional planning efforts. But it was also such coordinated economic planning by the regime that turned policy away from displaced farm workers to the automobile industry for answers. This turn toward heavy industry, and the accompanying economic and environmental concessions required by international corporations, is seen by many citizens as the final departure from the lightness of the first era of creative planning in the city and the final acceptance of a form of globalization in which citizens would suffer. It is not surprising or disappointing to some of them that the trajectory of planning and politics in the city brought about a serious political challenge to the Lerner regime from the Left.

There is a difference, then, between the reality seen by clients, which is static, and that seen by citizens, which is dynamic or, better, developmental. The citizens who lent their experience to this planning history, however, all agreed with Rodolfo Ramina about one thing: "There is no such thing as public participation in Curitiba." Although Ramina is aware that foreigners have reported on the so-called decentralization of government in the city's third planning period, he argues with considerable wryness that the Lerner regime "only decentralized paper work, there was no yielding of power to the neighborhoods." This is to say, as I argued earlier, that the clientelistic system of "exchanging favors" in Curitiba has kept citizens and citizenship weak but politicians strong.

Curitibanos, however, have a short distance to look for an alternative model of citizenship—to Porto Alegre, capital of the state of Rio Grande do Sul just to the south of Paraná. Although Paraná and Rio Grande do Sul share many cultural and ecological similarities, they could not have more different political dispositions.

The Challenge from the Left

Porto Alegre was governed by the PT (*Partido dos Trabalhadores*, or Workers' Party) from 1988 until 2004. In that brief six-year period, the city developed a planning culture no less unique than that of Curitiba, but one based on citizen participation in the annual budgeting process. Even when the PT unexpectedly lost the 2004 election, the participatory budgeting process was continued by the newly elected coalition. In simple terms, the city is divided into sixteen neighborhoods in which citizens participate in a nine-month-long conversation about how to prioritize the always-limited municipal budget. In the three remaining months of the year, elected representatives from each neighborhood negotiate their differing recommendations and finalize a budget before the cycle begins all over again. In 1999 some 40,000 people from a population of 1.3 million participated directly in the process. In contrast to decisions made by a few elite planners in Curitiba, the citizens of Porto Alegre rejected, for example, a proposed Ford automobile factory in the budgeting process because they felt the economic subsidy demanded by the company could be better spent elsewhere. This kind of decision by citizens can happen, observers argue, only as a result of solidarity gained through productive public talk (Goldsmith 2000).[1]

Although it would be a mistake to apply the experience of Porto Alegre directly to any other city, I will argue here that the PT has done more than any other political group in Curitiba, as in Porto Alegre, to replace clients with citizens. Some readers will find it paradoxical because the PT began life with explicitly Marxist, and thus undemocratic, credentials. The multiparty system of Brazil is indeed complex and filled more with big personalities than strict ideologies. The PT, however, is the exception rather than the rule. The lumber dealer Zanetti, for example, characterizes the PSDB—the party of former President Fernando Henrique Cardoso—as "a company party, not a national party." He means by this that Cardoso and his allies represent a chain of clients or overlapping interest groups, not a set of principles. In contrast the PT certainly has personalities, but these personalities agree on certain political ideals and methods of achieving them. In this sense, the PT is a political party that is similar to the EU model rather than its Brazilian counterparts.

The PT was founded in 1980 as a national party but with the strongest ties to the union movement of Sao Paulo. As a Marxist upstart, its ability to influence Brazilian public policy in the military era of government or even talk with those on the Right, let alone influential foreign governments or the World Bank, was severely limited. The role of the PT in that period was focused entirely on opposition to the military regime rather than on power. After Brazil's turn toward democracy began in 1985, however, the PT followed suit—it, too, declared adherence to the principles of policies of democracy (Ferrari 2001). The primary method the PT employed to move from its fully socialist origins to a center-Left position was the sometimes slow process of coalition building. Having participated in the successful ousting of the military, the PT focused first on the escalation of globalization and its implications for Brazilian workers, and, second, they focused on the crafting of regional and municipal coalitions founded on local issues. This conceptual strategy was what I will call mesopolitics—the desire to understand and resist those macroglobal forces that are at work in the world (corporate capitalism), but also the desire to understand and build microcoalitions—based on the reality of everyday life in a particular place. It is this view from the middle that distinguishes the PT from both the traditional macroview of Marxists and the always private interests of capital.

Nationally the PT won the presidency in 2002, and in the 2004 midterm elections won governorships in six states as well as 187 mayoral races throughout the country. In the eyes of many observers, this degree of success in such a short period of time polarized the politics of the nation, and those of Curitiba, between supporters of the PT and those of the PSDB. It will come as no surprise to some readers that in the early 1990s Jaime Lerner left his long-time party affiliation with the PDT (the Brazilian Labor Party created by Getúlio Vargas in 1945, which has been dominated by Leonel Brizola since 1979) and affiliated himself with President Cardoso and the PSDB. It is fair to say, then, that the confrontation between the PT and the PSDB in Curitiba was a microcosm of the nation as a whole.

In Curitiba, Angelo Vanhoni ran for the third time in 2004 as the PT candidate for mayor against Lerner's associate, Beto Richa. Without referring directly to the politics of Porto Alegre, Vanhoni campaigned on his commitment to "promote human development"—rather than economic development—and "involve the community through a style of management that stimulates public participation and engages civil society in the shaping of public policy" (Vanhoni 2004). Ederson Zanetti, among many others, assumed that this time Vanhoni would win because he was so far ahead in the polls just a few weeks before the election. That prediction, however, failed to account for the brutal—or brilliant, depending on your perspective—last-minute attack that the PSDB and the Lerner regime mounted against the PT. According to Zanetti and others, the Lerner regime illegally employed state funds in their "red scare" tactics, warning historically conservative Curitibanos that they would lose everything to "the communists." By associating the PT with the more radical Landless Worker Movement (MST, or Movimento dos Trabalhadores Rurais Sem Terra), Lerner was again able to redescribe the situation. It worked—Richa won the election by a margin of 4 percent (EIU 2004).

My purpose in recounting this story is to emphasize an argument made by Paulo Pereira. Since the presidency was won by the PT in 2002, citizen participation has become a federally mandated process in all regions and at all levels of government. Pereira holds that it has been implemented nearly everywhere, to one degree or another, except Curitiba. Although the Lerner regime has adopted a new slogan—Curitiba, "the social capital" of Brazil—most citizens understand the regime's new vocabulary as the reluctant acceptance of democratization forced on it by public talk initiated by the PT.

3.4 CURITIBA'S DOMINANT STORY LINE—TECHNOCRACY

Both clients and citizens, as well as foreign and Brazilian observers of the Lerner regime recognize that Curitiba's success is a function of its political continuity. Foreign clients of the regime, however, generally fail to ask how such political continuity was constructed. By contrast, insiders understand by experience—by their situatedness—that Curitiba's regime of sustainability is a clientelistic political system. This means that, in the frank words of senior IPPUC planner Celia Perez, "Each new mayor is hand-picked by his predecessor." If the chosen successor is previously unknown to the voters, his inconvenient anonymity is reconstructed in the media by identification with various IPPUC-sponsored public works projects (Meireles 1996). Until the emergence of the PT, elections in Curitiba were less about ideas than they were "plebiscites on the question of continuing a certain group in power" (Andrade 1996). As an affiliate of the regime, Perez argues that such undemocratic continuity is "good for us."

For North Americans, who have enjoyed political continuity and stability for so long, it is easy to underappreciate Perez's perspective. After a long period of political and economic instability it is understandable that many Brazilians would welcome a social order that demonstrates itself to be progressive even if secretive and elitist. Many readers will have noticed by now, however, that it is only insiders affiliated with the Lerner regime—its clients—who credit undemocratic processes as being in their interest.

In sum, it is fair to characterize the dominant story line constructed by the Lerner regime as *technocracy*. This term is rather easy to pick apart—it is obviously related to the modern words *techno*logy and demo*cracy*.

Without conducting a formal etymology from the Greek origins of these words, we can intuit a meaning that suggests something like "technological rule" or "rule by means of technology." This is not as far-fetched an idea as it might seem at first glance. Several contemporary philosophers, Langdon Winner and Andrew Feenberg principal among them, argue that this hybrid word—*technocracy*—reflects the way we actually live. In Winner's view, choosing a technology is not choosing an artifact—a computer, for example—it is choosing a way of life. Purchasing the computer commits you to hours of e-mailing every day and the purchase of antivirus software, creates access to new products you cannot touch before purchasing them, and so on. If we extend this logic, the term *technocracy* suggests only that the person who chooses our technologies chooses our way of life (Winner 1999; Feenberg 1999).

In common parlance, however, technocracy usually refers to the domination of government by a cadre of technically trained experts whose interests are limited to restricting public choices to the most efficient, not the most pleasant, most beautiful, or most sustainable. In the case of Curitiba, the experts who are charged with making decisions about how citizens live are not technocrats in the most restrictive sense of making only efficient choices, but neither are they agents of public participation. Somewhere in this range of possibilities lays Curitiba's technocratic regime of sustainability.

As I did for Austin, I will characterize Curitiba's dominant story line as being constituted of three kinds of public talk: political, environmental, and technological—that are both local and global in scope, adding up to a version of reality that is shared among those who see the world as desperately in need of an authoritative power structure that will discipline clients, the marketplace, and the natural world. The technocratic story line is summarized in table 3.1.

Curitiba's Dominant Political Talk

The political talk that coheres with a technocratic story line is what Barber (1984) terms *liberal-realism*. Liberal-realists are part of the Western democratic tradition because their intentions are benevolent, if—as McKibben recognized—paternalistic. Their generic response toward social conflict is to suppress it because progress toward sustainability depends on the fragile stability they are mandated to maintain. Toward that end, liberal-realists depend on empirical evidence to develop predictable "action-scripts," which are expressed as official "rules of thumb" that are employed to design and implement solutions to problems. In turn, acknowledgment of the wisdom behind the rules of thumb legitimates the regime's power, and power is required because people need discipline, not license. Public

Table 3.1. The Dominant Story Line of Curitiba—Technocracy

Kinds of talk	As told by: Celia Benato, Paul Hawken, Josef Leitman, Jaime Lerner, Amory Lovins, Hunter Lovins, Tony Lloyd-Jones, Bill McKibben, Celia Perez, and Jonas Rabinovitch
Dominant Political Talk	*Liberal-Realism*
Generic response to conflict	Suppression— Conflict disrupts the stability upon which progress depends
Attitude toward certainty	Empirical— In pursuit of absolute power
What is valued	Wisdom— Which legitimates power
Conceptualization of space	Restricted— Individuals necessarily exist in zoolike environments where conflict is suppressed by zoo keepers
Dominant Environmental Talk	*Administrative Rationalism*
Basic entities recognized or constructed	Experts, managers, liberal capitalism, and the administrative state
Assumptions about natural relationships	Nature is subordinated to human problem solving and citizens are subordinated to the administrative state, which is managed by experts
Agents and their motivation	Only experts have agency, but they are motivated by the public interest
Key metaphors and rhetorical devices	Administrative creativity, reassurance, and concern
Dominant Technological Talk	*Technological Display*
Attitude toward technology	Technophilic— Technology is good, when correctly designed, because it increases productivity and efficiency while it minimizes environmental costs
Attitude toward history	Contingent— The future depends upon wise choices demand being made by experts that will satisfy client
Assumptions about the origins of technology	Collaborative genius— Creative interdisciplinary groups of experts invent the landscapes that society needs to prosper
Types of tools employed	Incremental— Those that will produce maximum aesthetic and material affect at least cost

space, then, is conceptualized as restricted—necessarily policed for the benefit of all.

Curitiba's Dominant Environmental Talk

The environmental talk that coheres with a technocratic story line is what John Dryzek (1997) terms *administrative rationalism*. What technocrats recognize as legitimate entities includes resources, capital, experts, and the administrative state that manages them. In this view of the world, nature is subordinated to human problem solving; citizen-clients are subordinated to the state that is, in turn, administered by experts. As a result, only experts have agency, but they are motivated by the public interest (most of the time). Key metaphors employed by administrative rationalists include "creativity" and various terms designed to reassure citizen-clients and demonstrate the concern of the regime for their well-being.

Curitiba's Dominant Technological Talk

The technological talk that coheres with the technocratic story line is what I call *technological display*. In this view of the world technology is good, when correctly designed, because it increases human productivity and minimizes externalities or the unintended environmental costs that can come back to haunt stability in the future. The viability of the future, then, depends on experts making wise choices that will satisfy public demand in both the short and long term. Experts, however, are not individual geniuses working in isolation, but rather teams of creative collaborators who design the technological landscapes that society needs in order to prosper. In an emergent economy like Brazil's, the types of tools most needed are those that will produce maximum aesthetic and material affect at least cost. Lerner has argued, "You have to understand you are responsible for the hope of the people, . . . if the city isn't changing, then you're frustrating their hope" (cited in McKibben 1995). It is, then, the aesthetic content of technology that raises public hope. Technological displays such as these create collective experiences that transform the public perception of reality (Nye 1994).

3.5 CURITIBA'S COUNTER-STORY LINE—SOCIAL DEMOCRACY

Like many PT supporters, planner Celia Benato retells the PT story line as an attempt to construct a form of social democracy that North Americans tend to associate with European Union rather than South American governments. This hopeful speculation is summarized in table 3.2. Although

Table 3.2. The Counter-Story Line of Curitiba—Social Democracy

Kinds of counter-talk	As told by: Roberto Allende, Celia Benato, Márcio de Oliveira, Paulo Pereira, and Rodolfo Ramina
Political counter-talk	*Citizen participation*
Generic response to conflict	Transformation— Conflict creates an opportunity for change
Attitude toward certainty	Developmental— Reality is always emergent
What is valued	Participation— Stability and security require strong citizenship
Conceptualization of space	Continuous— Individuals exist within the social fabric
Environmental counter-talk	*Green Rationalism*
Basic entities recognized or constructed	Global forces (corporate capitalism), local practices, and natural resources
Assumptions about natural relationships	Equality should exist between all people. The food security of humans is directly linked to ecological integrity
Agents and their motivation	There are many agents with multiple motivations, but the agency of nature itself is downplayed
Key metaphors and rhetorical devices	Social (not economic) development and environmental conservation
Technological counter-talk	*Ad Hoc-ism*
Attitude toward technology	Skeptical— Neither naively technophilic nor technophobic. Technologies contain politics and must be evaluated on a case-by-case basis
Attitude toward history	Both teleological and contingent— Some hold on to Marxist ideas about historical inevitability, and others argue a brighter future depends upon the political will of citizens to craft just tools
Assumptions about the origins of technology	Both deterministic and socially constructed— Some hold on to Marxist ideas that technology determines history, and others argue that technologies emerge from competing interests and values
Types of tools employed	Both high-tech and appropriate— Technological choices should be made that balance labor and capital input to desired output

the PT story line is not the strong form of democracy proposed by Barber (1984), it bears certain similarities.

Curitiba's Counter-Political Talk

The political talk engaged in by Curitibano citizens, if not clients of the regime, is mostly about *citizen participation*. The generic response to conflict by social democrats is not to suppress it but to welcome it as an opportunity for change. Because the immigrant workers of Curitiba are largely invisible to elites, conflict creates not only visibility but also an opportunity to transform existing conditions through participation, which is valued above the certainty of outcomes. And if participation is valued, space is conceptualized differently. The *favelas*, where many landless workers reside, are experienced as an informal, flexible, and organic social fabric rather than adjoined private spaces.

Curitiba's Counter-Environmental Talk

The environmental talk that coheres with the story line of social democracy is what Dryzek (1997) refers to as "green rationalism." What social democrats recognize as being "out there" is a globalized economy in which corporate agribusiness has emptied the countryside of local practices and exploited natural resources in which corporate manufacturers vie for the lowest labor and material costs. As former Marxists, the green rationalists of Curitiba argue that equity should exist between all people but that the food security of humans is linked to ecological integrity. This is clearly an anthropocentric view of nature, but one that links environmental conservation and social development.

Curitiba's Counter-Technological Talk

The technological talk that coheres with the story line of social democracy in Curitiba is, like that of the Lerner regime, ad hoc. PT candidates openly admire the "small is beautiful" era of modern planning in the city and propose to use similar rationality but in the pursuit of different goals (Vanhoni 2004). And, like Lerner, they are aware that technologies carry politics and must be tested on a case-by-case basis. They are neither technophiles nor technophobes. If, however, there is a fairly consistent attitude toward technology being a social phenomenon, there are mixed attitudes toward its origins and role in history making. The wide range of attitudes I found among PT supporters and sympathetic citizens suggest that some hold traditional Marxist attitudes toward history and technology whereas others hold views consistent with the liberal capitalist views of the Lerner regime. Supporters of the MST, for example, promote the use of "appropriate technologies," but others promote very high-tech digital tools. On the basis of such a small sample, I can conclude only that the process of coalition building makes for strange ideological bedfellows when it comes to choosing tools. Social democracy in Curitiba, then, is talk about technology with ad hoc fluidity.

3.6 SUMMARY

Bill McKibben (1995, 79) argued in the mid-1990s that "Lerner—and Curitiba—seem to have shed ideology in the name of constructive pragmatism." On his account we should consider Lerner and his colleagues to be *pragmatists*. On first glance, McKibben's claim might be written off as a casual reference to the popular or vulgar variety of pragmatism, not that which derives from the texts of Peirce, James, and Dewey. In the popular view, you are a pragmatist if you would subscribe to the slogan "whatever works" to rationalize the means employed to achieve a desired outcome. This is clearly not what the philosophers had in mind. But neither did Lerner. In fact, it now seems reasonable to argue that the pragmatists and Lerner do share an antipathy for what we can call "first principles"—those absolute claims to historical truth that are made by theorists on both the Right and the Left of the political spectrum. They also share a common interest in methods of implementation and outcomes. But, if there is common ground between the pragmatists and Curitiba's regime of sustainability, there is also a chasm of disagreement. If the pragmatists have anything that approaches the stature of a first principle it is their belief in democratic process and the public talk that is routinely suppressed in the efficient technocracy of Curitiba. We are left with a dilemma that requires further analysis to determine whether there is a kernel of truth in McKibben's characterization of the Lerner regime.

In light of the centrality of Jaime Lerner in both versions of this story as either hero or villain, it seems necessary to consider the role leaders play in the production and maintenance of democracy and political talk. The question of leadership is indeed a difficult one and it is naïve to hold that effective democracy requires no leadership at all. Barber (1984, 241) holds that charismatic leadership—the kind furnished by Winston Churchill, John F. Kennedy, or Lerner—is a "substitute for the ability of polity to make autonomous choices," not the enabler of informed public choice. This assessment appears entirely consistent with the conditions in Curitiba. If Barber is correct, it suggests that the Lerner regime has not enabled sustainable development beyond the political lifetime of the regime itself. After examining two cities, though, it is hard to argue that

Austin's regime of sustainability has been more successful than Lerner's. In fact, the technocrats of Curitiba can legitimately claim to have enabled as much environmental sustainable development as have the environmentalists of Austin. This assessment begs the question: might such an authoritarian substitution be better than no informed choice at all?

Four North American Assessments

Thomas Prugh and his coauthors argue that "it is possible (barely) to imagine that an enlightened despot might come to power somewhere and put his country on the road to sustainability by edict" (Prugh 2000). But, although authoritarian forms of governance might develop the economy, preserve the environment, and improve social equity, there are several reasons why participatory forms of governance are more likely to do so.

First, elites can, and do, insulate themselves from problems, particularly those related to environmental degradation. It is unimaginable, for instance, that the designer-politicians of IPPUC (or PT officials) would live in the many *favelas* of the city. Second, problems requiring collective action to correct are more effectively resolved by democratic horizontal networks rather than by autocratic vertical networks. Simply put, information and solidarity do not easily flow uphill. And third, but most significant, technocratic elites tend to suppress conflict in order to hold onto power in the name of social stability. What technocratic regimes fail to understand is that conflict is the very social dynamic that provides the feedback loops necessary for problem identification and resolution. By suppressing conflict, technocracies suppress the very information they need most to govern wisely, and, if the reader will recall my earlier claim, "wisdom" is the social value on which technocratic authority is legitimated. Political legitimation based on superior wisdom is then short-circuited by the exercise of power itself.

Sympathetic arguments—all of which favor the democratic construction of sustainable environments—are made by three other North American authors: Bernie Jones, William Shutkin, and Robert J. Brulle. Jones argues for citizen participation and against the professionalization of urban planning, because, first, citizen-made plans will more accurately reflect their needs and concerns than will the guesses of experts. Jones holds that local rather than expert knowledge is essential to finding effective solutions. Second, the more public participation in an issue, the more citizens will insist on implementation. Third, the more citizens participate the harder it is for officials to ignore the plan (Jones 1990).

By use of related logic, Shutkin argues against the proliferation of "professionalism" among environmental organizations in North America. He holds that professionalism reflects the ill-informed nineteenth-century idea that power within society is commensurate with one's accumulated, nontransferable knowledge. The development of expert cultures, then, has not only come "at high cost to democracy" but has also been largely ineffective in preserving threatened natural environments. This finding suggests that there is a direct correlation between environmental degradation and the amount of participation and planning on the part of the diverse stakeholders in any specific situation. In other words, environmental degradation is simply an indication that the interests of some stakeholders are being ignored (Shutkin 2000).

The fourth North American author, Brulle, employs a Habermasian perspective in favoring democracy over enlightened despotism. He argues that because both the market and the state have demonstrated limited capacity for self-correction regarding socially created environmental degradation, social learning must come from outside these institutions. In his view, environmental problems are social rather than technological in character—a perspective that evades most technocrats because they legitimate their social power on the basis of technological expertise. As a result, "democracy is a key component in enabling . . . social change." Because "social learning is based upon open communicative action in the lifeworld, the enhancement of the democratic will-formation capacity of society is a necessary prerequisite for the initiation of the actions through which environmental problems could be achieved" (Brulle 2000, 64–65).

In sum, these four North American authors allow for the existence of an authoritarian version of sustainability—the kind Barber terms "liberal-minimalism." In his words:

> Democracy in the authoritarian mode resolves conflict . . . through deferring to a representative executive elite that employs authority (power plus wisdom) in pursuit of the aggregate interests of its electoral constituency. (Barber 1984, 140)

Although all five authors allow for the existence of an authoritarian version of sustainability, they present convincing arguments that favor the long-term success of a strongly democratic version. On these grounds, along with my own analysis of local conditions, it is reasonable to argue that strongly democratic regimes will more successfully enable sustainable development than will technocratic ones. But, as I argued at the beginning of this chapter, it is certainly a mistake to assess Curitiba's regime of sustainability through North American or European cultural assumptions alone. Such an assessment not only would be unfair but would also likely miss what there is to learn from Curitibanos.

Looking Forward

What, then, can we finally say about Curitiba? First, that it is technocratic and a good exemplar of Barber's liberal-realist politics. But second, it is also a city like Austin in that is "in the making." What I mean by this that it would be premature to declare the city's reputation to be little more than propaganda. Rather, it is necessary to acknowledge what has been accomplished there—not just the production of statistics concerning reduced fuel consumption, increased green space per capita, or narrowed income distribution, but the sense among Curitibanos that their city is a political—rather than material—construction project. If the regime of sustainability is, by design or by accident, starting its citizens down the road toward a strong, rather than weak model of citizenship, sustainable development is *becoming* true.

Third, we can say that Curitiba is already efficient by almost any standard. But, as Bill McDonough has argued, efficiency in a philosophical sense "has no independent value." Its value is, rather, dependent on the larger system of which it is a part. "An efficient Nazi, for example, is a terrifying thing" (McDonough 2002, 65).

My historical analysis of Curitiba suggests that significant change is already underway and that, for practical purposes, the Lerner regime of sustainability is now a historical phenomenon. In the eyes of locals the city has passed out of its heroic period of infrastructure development fostered by the Lerner regime and into a new period. The change is, in part, due to the emergence of a new generation of leaders and citizens who are also participants in the counter-story line of social democracy that has emerged nationwide. Although it is too early to know how the presidency of Lula de Silva will push history in Brazil, there is already some sense that the dominant and counter-story lines of the city are merging. And as story lines intersect, deflect, and evolve, new alliances and new regimes will be constructed.

Lessons Learned

If it is a mistake to assess the Lerner regime of sustainability only through North American or European lenses, the same logic works in reverse—it would be a mistake to directly appropriate Curitiba's planning methods because the cultural and political context of that city is so very different than anywhere else on the planet. This is not to argue, however, that there is nothing to learn from the experience of Curitiba. Quite to the contrary, I find three lessons to be learned that take us back to the odd affinities between philosophical pragmatism and Curitiba's technocratic planning culture. These are what I will refer to as *abductive logic*, the *theory of implementation*, and *incrementalism*.

Hilda Blanco (Blanco 1994) credits the Lerner regime with employing abductive reasoning in planning and offers a concrete example to illustrate how abduction differs from either deductive or inductive reasoning, which are the cultural norms of North America and Europe. When confronted with the need to extend city infrastructure (water and sewer services) into informally settled areas, or *favelas*, planners were aware that if they also extended paved streets, which is the normal engineering practice, land prices would rapidly escalate and drive residents to another sector of the city without such services. In other words, engineering best practices (derived from inductive reasoning) would have consequences that were contrary to the social goals of providing infrastructure in the first place, which was to improve public health and thus decrease municipal health care costs. Alternately, the use of deductive reasoning to solve this problem would suggest that planners could, for example, act on the basis of the trickle-down theory that would accept the dislocation of residents on the assumption that wealth and good health work their way down the social order. What planners actually did, however, was to develop an unorthodox technological solution to an agreed-on social goal—they extended sewer and water lines along existing foot paths but did not construct vehicular streets. The consequence was that the *favelas* remained socially stable *and* became more healthful places to live.

The lesson to be learned here—one that may be transferable—is that technological norms in any culture bring with them political consequences that may or may not be desirable but are generally hidden. What distinguishes abductive rationality from engineering best practice is that it employs a situated perspective of the problem at hand rather than an abstract one. This is a distinction that will be discussed at greater length in chapter 6. I open the topic here because abduction was recognized as a legitimate and oft preferred form of rationality not only by Aristotle but also by the American pragmatist Charles Sanders Peirce. This seeming coincidence provides yet another commonality between the planning culture of Curitiba and philosophical pragmatism.

The second lesson learned is consistent to a degree with Schwartz's analysis. In his final assessment of Curitiba, Schwartz (2004, 31, 86, 116) argues that the city's unexpected urban renewal success was not the result of "urban planning theory," sophisticated "cost-benefit analysis," or even "traditional economic analysis." Instead, he argues, the Lerner regime had organically developed an informal "theory of implementation" that apparently derived from the culture of architectural problem solving in which regime leaders had been trained. This is indeed a surprising finding by a scholar trained in the neoclassical economic tradition. Perhaps even more surprising is that Schwartz's term, *theory of implementation,* might stand as a rather good characterization of philosophical pragmatism.

Larry Hickman has proposed that pragmatism, though not a "philosophy of action," can be characterized as a "philosophy of production" in which "the goal of inquiry is not action, but the construction of new and more refined habits, tools, goals, and meanings, in short, new and more refined products" (Hickman 2001). Although the language employed by Schwarz and Hickman is not identical, their intended meanings are, I think, clearly related. For both authors, as for Curitiba's planners and pragmatists in general, a theory of implementation will always be more important than the conceptualization of goals. This claim rests on the empirical knowledge that no matter how one conceptualizes a problem, one's understanding will evolve as one attempts to solve it. This is to say that successful implementation is effective reconceptualization, or methods of implementation shape outcomes more than initial goals.

The third lesson learned from Curitiba's planning culture may be a corollary of the first two. Although most observers have credited the Lerner regime with planning comprehensively, I argued above that IPPUC succeeded because it planned incrementally through what Rabinovitch has called "action scripts." The great benefit of action scripts of short duration, rather than long-term comprehensive planning, is that these humble projects allow for almost immediate feedback that can then inform the next incremental project and so on. The implications of this strategy are very significant and will be discussed further in chapter 6. For the moment it will suffice to say that incremental planning is not the absence of planning theory, but a theory of implementation that is skeptical of certainty.

Finally, I am left with the conclusion that McKibben's casual characterization of Curitiba's planning culture as "pragmatist" has more than an ounce of validity. The problem, as I acknowledged earlier, is that pragmatism without democracy is a bit like Marxism without history—it is logically incoherent. This is not to say that the planning culture of Curitiba *must* move in the direction of democratic inclusion. Not at all. Philosophies require logical coherence; people and cultures do not. As we have seen earlier, the story lines of any culture are invariably contradictory. It is to say, however, that there is an affinity between some stories and practices found in Curitiba and those promoted by philosophical pragmatism. In affinity there is opportunity.

CHAPTER REFERENCES

Andrade, P. (1996). "Brazil: Election results in six capitals analyzed." <wnc.fedworld.gov/> (accessed 9 October 2001).

Barber, Benjamin. (1984). *Strong democracy: Participatory politics for a new age*. Berkeley: University of California Press.
Blanco, H. (1994). *How to think about social problems: American pragmatism and the idea of planning*. Westport, CT: Greenwood Press.
Boles, D. (1992). "Brazil's modest miracle." In *Landscape Architecture* 82 (6): 58–59.
Brulle, R. J. (2000). *Agency, democracy and nature: The U.S. environmental movement from a critical theory perspective*. Cambridge, MA: MIT Press.
Cavacanti, C. (1995). "Brazil's urban laboratory takes the strain." In *People & the Planet* p. 6. <www.oneworld.org/> (accessed 7 February 2000).
Cervero, R. (1995). "Creating a linear city with a surface metro." In *Faculty Research, Institute of Urban and Regional Development*. University of California at Berkeley. <www-iurd.ced.berkeley.edu/> (accessed 22 March 2001).
de Oliveira, M. (2001). "A trajetoria do discurso ambiental em curitiba (1960 2000)." *Revista de Sociologia e Política* (16): 97–106.
Di Giulio, S. (1994). "Architect, mayor, environmentalist: An interview with Jaime Lerner." *Progressive Architecture* 75 (7): 84–85, 110.
Dryzek, John S. (1997). *The politics of the earth: Environmental discourses*. Oxford and New York: Oxford University Press.
Editorial. (1997a). *The State of Sao Paulo*, p. A3. <www.samek.com.br/> (accessed 22 April 2001).
Editorial. (1997b). "Leading Brazilian business," December 9. <www.samek.com.br> (accessed 12 March 2001).
EIU. (2004). "Richa wins in Curitiba." In *Economist*. EIU.com (accessed 12 September 2005).
Feenberg, A. (1999). *Questioning Technology*. London: Routledge.
Ferrari, A. (2001). "Testing times for the workers' party." *Socialism Today: The monthly journal of the workers party*. Issue 55 (April). <www.socialismtoday.org/55/brazil.html> (accessed 16 September 2005).
Filho, A. (1997). "Brazil: LBGE reveals population's changing face." In *O Globo: World News Connection*. <wnc.fedworld.gov/> (accessed 2 April 2001).
Fischer, F. (2003). *Reframing public policy: Discursive politics and deliberative practices*. New York: Oxford University Press.
Goldsmith, W. (2000). "Participatory budgeting in Brazil." In *Planners' Network*. <www.plannersnetwork.org/resources/pdfs/brazil_goldsmith.pdf> (accessed 16 August 2004).
Hawken, P. L., Amory Lovins, and Hunter Lovins. (1999). *Natural capitalism: Creating the next industrial revolution*. Boston: Little, Brown.
Hess, D. and Roberto A. Damatta. (1995). *The Brazilian puzzle: Culture on the borderlands of the western world*. New York: Columbia University Press.
Hickman, L. A. (2001). *Philosophical tools for a technological culture: Putting pragmatism to work*. Bloomington: Indiana University Press.
Hunt, J. (1994). "The Urban Believer: A report on Jaime Lerner and the rise of Curitiba, Brazil." In *Metropolis Magazine* 13 (8): 66–67, 74–77, 79.
Irazabál, C. (2005). *City making and urban governance in the Americas: Curitiba and Portland*. Hants, England: Ashgate.
Jones, B. (1990). *Neighborhood planning: A guide for citizens and planners*. Chicago: Planners Press.

Kant de Lima, R. (1995). "Bureaucratic rationality in Brazil and the United States: Criminal justice systems in comparative perspective." Pp. 241–69 in E. David Hess and Roberto Damatta, eds., *The Brazilian Puzzle*. New York: Columbia University Press.

Katz, I. (1997). "Building a Detroit in Latin America." In *Business Week*, September 15, p. 58.

Lerner, J. (1978). *Urban development in Brazil*. Transcript of lecture delivered March 28 at Salvador, Brazil. Collection of the author, University of Texas.

Lerner, J. (1996). "Change comes from the cities." In *Human Settlements*. United Nations Centre for Housing, Building, and Planning, Dept. of Economic and Social Affairs. <www.un.org/esa/publications.html> (accessed 7 June).

Light, A. and E. Katz, eds. (1996). *Environmental pragmatism*. London: Routledge.

Linden, E. (1996). "The exploding cities of the developing world." In *Foreign Affairs* 75, pp. 52–65.

Lloyd-Jones, T. (1996). "Curitiba: Sustainability by design," *Urban Design Quarterly* (January). <rudi.herts.ac.uk/ej/udq/57/csd.html> (accessed 10 February 2001).

Mang, R. (1995). "Principles for sustainability." In *Community building: Renewing spirit and learning in business*. New Leaders Press. <www.vision-nest.com/cbw/Principles.html> (accessed 10 October 2004).

McDonough, W. and M. Braungart. (2002). *Cradle to cradle: Remaking the way we make things*. New York: North Point Press.

McKibben, B. (1995). "Curitiba." In *Hope, human and wild*. Boston: Little Brown.

Meadows, D. (1995). "The city of first priorities." In *Whole Earth Review* 85 (2) (Spring 1995): 58–59.

Meireles, A. (1996). "With the blessing of the godfathers." In *ISTOE: World News Connection*, Sao Paulo. August 21. <wnc.fedworld.gov/> (accessed 21 April 2001).

Menezes, C. L. (1996). *Desenvolvimento urbano e meio ambiente: A experiência de curitiba*. Campinas, Brazil: Papirus.

Moura, R. and C. Ultramari. (1994). *Grande Curitiba: Teoria e práctica*. Curitiba, Brazil: Ipardes.

MST. (2004). *Movimento dos trabalhadores rurais sem terra*. <www.mst.org.br/> (accessed 7 March 2001).

Nye, D. (1994). *American technological sublime*. Cambridge, MA: MIT Press.

Prugh, T., R. Costanza, and H. Daly. (2000). *The local politics of global sustainability*. Washington, D.C.: Island Press.

Rabinovitch, J. and J. Lietman. (1996). "Urban planning in Curitiba." *Scientific American* 274 (3) (March): 46–53.

Ramina, Rodolfo. (2005). Telephone interview, 28 February.

Richards, J. (1997). "Jorge Wilhelm: Desiring global change." In *Architectural Design* 67 (1) (January–February): 14–17.

Rorty, R. (1998). *Achieving Our Country*. Cambridge, MA: Harvard University Press.

Samek, J. (2001). "Governador manda suspender ped." <www.samek.com.br/> (accessed 1 December).

Schwartz, H. (2004). *Urban renewal, municipal revitalization: The case of Curitiba, Brazil*. Alexandria, VA: Hugh Schwartz, Ph.D.

Segwaw, H. (1998). "Alternative bureaucracy in Curitiba." In *Spazio e Societa* (83) (April–June): 40–52.
Shutkin, W. (2000). *The land that could be: Environmentalism and democracy in the twenty-first century*. Cambridge, MA: MIT Press.
Skidmore, T. E. (1967). *Politics in Brazil, 1930–1964: An experiment in democracy*. New York: Oxford University Press.
Suplicy, M. (2003). "Globalization and exclusion." In *City of São Paulo publications*. <see.oneworld.net/article/search/> (accessed 17 October 2005).
Vanhoni, A. (2004). "An interview with Angelo Vanhoni." <www.fpa.org.br/noticias/entrevista_vanhoni.htm> (accessed 7 July 2005).
Vassoler, I. (2003). "Urban visions: Lessons in governance from two Brazilian cities." In *Proceedings of the twenty-fourth Annual Conference of the Middle Atlantic Council for Latin American Studies*, February 21–22, 2003 at Kutztown University, Kutztown, PA. *MACLAS XVII*, p. 17. <www.maclas.vcu.edu/journal/Vol%20XVII/index_XVII.htm> (accessed 17 October 2004).
Winner, L. (1999). "Do artifacts have politics?" Pp. 28–41 in D. MacKenzie and J. Wajcman, eds., *The social shaping of technology*. Philadelphia: Open University Press.
Worcman, N. (1993). "Boom and bus." *Technology Review* 96 (8): 12–13.
Xavier, V. (1975). *Histõria de curitiba em quadrinhos; boletim informativo no. 11*. Curitiba, Brazil: Fundação Cultural de Curitiba.

NOTE

1. This claim by Goldsmith is vigorously contested by Professor Benamy Turkeinicz, of the Universidade Federal do Rio Grande do Sul, who argues that Ford was lured away by a better offer elsewhere, not rejected by citizens. Turkienicz does not, however, contest the more fundamental observation made by Goldsmith: that the participatory taxation process of Porto Alegre is a strongly democratic one. The conflict on this point is yet another example of the differing perspectives of insiders and outsiders.

FOUR

The Banks of Frankfurt

Many were surprised when Volker Hauff was elected mayor of Frankfurt in 1989. As one of the twenty official authors of the Brundtland Report, *Our Common Future* (WCED 1987), he was a prominent advocate of environmental sustainability. Coming on the heels of a municipal era dominated by the right-of-center Christian Democratic Union (CDU), his election marked a clear shift in the city's politics. Hauff's election was not a personal victory based on populist appeal as was the case in Curitiba. Rather, it was made possible by a newly forged coalition between the left-of-center Social Democratic Party (SPD) and the German Green Party—an association that Germans refer to the "red/green coalition." It was this homegrown coalition that became the city's regime of sustainability, and it lasted for six years, until 1995.[1] (See figure 4.1.)

The contemporary story lines of Frankfurt, like anywhere, are bound up in its history, but, because the history of this city is considerably longer and more tragic than that of Austin or Curitiba, it is far more tempting to suppress particular story lines in favor of others. As a result, there are tensions between contemporary and deeper accounts of Frankfurt's story that deserve attention. Both accounts revolve around competing interpretations of Frankfurt's identity as a banking city, so my analysis will begin there. But before reconstructing these differing accounts of Frankfurt's banking culture, I should place them in the context of this investigation.

Readers may by now suspect that the order of these three urban stories is teleological. By this I mean that just as Curitiba has been more successful than Austin in achieving the elusive goal of sustainable urban development, so Frankfurt seems to have achieved more than Curitiba. The structure of my analysis, then, has proceeded from good to better. This pattern suggests that the political, environmental, and technological discourses constructed in Frankfurt add up to story lines that have most effectively engendered sustainable development. This is not to argue that Frankfurt is the best model of sustainable urban development in the world. Many Europeans argue that, even within the context of the European Union (EU), there are several other

Figure 4.1. Location Map of Frankfurt

cities that deserve equal, if not more, attention. For my purposes, however, Frankfurt offers a story that is particularly good in helping us to better understand the dynamics of competing types of public talk.

The dominant story line of Frankfurt—what I refer to as *progressive capitalism*—is constituted of public talk that strives toward political tolerance, ecological modernization, and technological progress. These are values that the authors of the Brundtland Report saw as generically consistent with sustainable development as they defined it in 1987. In chapter 1 I argued that the Brundtland discourse defines sustainable development as the balancing of economic development, environmental protection, and social equity. In my analysis of Frankfurt, however, I find that one of these variables—economic development—is always more equal than the others. This should hardly be a surprise in a banking culture in which most choices are economically determined. What is of considerably more interest is how the values of economic determinism were subverted, even if briefly, by an opposing story line constituted of public talk that strives toward strong democracy, green development, and the social determination of technology.

The red/green story line told by Frankfurt's regime of sustainability brought to fruition a sometimes volatile thirty-year-long public conversation about the nature of urban life. By constructing a coalition that united the interests of those committed to environmental preservation and those committed to social equity, the red/green coalition was able to wrest power from the traditionally dominant coalition of business and real estate interests committed to economic development. The result was an urban form considered by many to be not entirely German. Many have argued that Frankfurt's skyline, now dominated by "skyscrapers," is American. (See figure 4.2.) I will argue, however, that it is Frankfurter. This story of redescription provides lessons related to coalition building, conflict resolution, economic determinism, and code making that is particularly helpful in our effort to understand alternative routes to the sustainable city. As in the first two case studies, however, a historical approach will be helpful.

4.1 TOLERANCE AND BANKING

The recent history of Frankfurt's banking industry begins, according to popular accounts, in 1949 with the narrow defeat of the city's nomination to become the capital of the new West German nation (Scherf 1998). It seems that West Germany's first chancellor, Konrad Adenauer, favored Bonn as a temporary capital, rather than Frankfurt as a permanent one. The loss of this distinction was particularly painful to Frankfurters because the city has had a long association with German aspiration to nationhood. In fact, until the Holy Roman Empire of the Germanic Nation ceased to exist in 1806, thirty-three of Germany's fifty-two kings and emperors were elected within the city's walls. In 1240, Frankfurt was granted imperial protection as a trade center, and in 1382 it officially became a free imperial city—meaning that the burghers of Frankfurt answered directly to the emperor. In more recent history Frankfurt was the capital of the German Federation from 1816 to 1866, and from 1848 to 1849 the *Paulskirche* in Frankfurt hosted the emergent nation's first democratically elected parliament. The post–World War II rejection of this distinguished political history, popular accounts claim, stimulated the imagination of citizens to fabricate an alternative, or consolation, identity for the city—that of the nation's financial capital. If we cannot become the nation's political capital, this logic

Figure 4.2. Frankfurt Skyline (2002).
Courtesy of the author.

argues, we will become its bank. Secondary evidence lends credibility to this popular interpretation of Frankfurt's modern transformation. On March 18 and 22, 1944, much of the historic center of Frankfurt was destroyed by Allied bombing raids. This tragically cleared real estate became a prime development opportunity in the postwar city. What would have been declared to be a historically significant, and thus protected, stock of half-timbered medieval houses became instead a development opportunity. Add to this situation Frankfurt's central geographic location in Europe, its hub airport, and railway network, and the postwar decision to move the German banking industry from Düsseldorf appears only rational. This popular account of Frankfurt's modern history is promoted by such mainstream publications as *Der Spiegel* (Schreiber 1997) and was relayed to the author by, among others, Greenpeace activist Jürgen Braune and Commerzbank executive Arnheim Müller.

This interpretation of Frankfurt's apparently resilient history, however, neglects uncomfortable information. The discomfort stems from an argument made by Frankfurt city planner Helmut Bosch. In his view, the city's primacy as a trading center—which dates back to the eleventh century—was enabled by the Franks' complex association with its repressed Jewish minority. Historically, Jews in Europe were assigned those economic functions deemed unsavory by premodern Christians but deemed functionally essential by emergent capitalist societies. In other words, it is hard to imagine how Frankfurt could have become the significant medieval trading center that it was without the economic incentive provided by the institutionalization of interest-bearing loans based on real collateral. It was these economic functions that were administered by Jews, and that enabled the city's sophisticated trade economy. This schizophrenic social relation between Franks and Jews was played out in alternate periods of persecution and tolerance.

It is well documented that Jews were minority residents in Frankfurt as early as the twelfth century. In 1349 the expanding Jewish population was assigned blame for an outbreak of the plague and mercilessly persecuted. A century later the traditional Jewish quarter near the cathedral was declared off limits to Jews, who were subsequently relocated en masse to a far smaller ghetto near the present-day Borneplatz known as the *Judengasse*, or Jew's Alley. In 1500, there were some 200 Jewish inhabitants of the ghetto that rapidly grew, by 1550, to be 1,000 of Frankfurt's total population of 12,000.

By the seventeenth century, Jewish bankers such as Samson Wertheimer had secured the support of Emperor Charles VI by helping him finance his military ambitions. The institutional relation of Jews to the emperor is best described as a condition of semifreedom in which imperial protection was granted as a condition of servitude to the imperial treasury. The rising visibility and aspirations of Jews, it is said, helped to provoke the peasant revolt of 1614 that again led to the plundering of the ghetto and the expulsion of Jews from the city. The emperor, however, intervened militarily on behalf of his bankers and forcefully repatriated Jews to Frankfurt in 1616. In this version of the story, Frankfurt became an independent imperial city, not on the account of Christian guilds and burghers but on the account of Jewish bankers.

In spite of such systemic persecution and devastating ghetto fires in 1711, 1721, and 1796, the Frankfurt Jewish community grew to be one of the most important centers of Jewish culture in seventeenth- and eighteenth-century Europe. It was this legacy of Jewish industriousness best exemplified by the career of Meyer Anschel Rothschild (1743–1812)—founder of the famous Rothschild Bank—that is responsible for much of Frankfurt's nineteenth-century economic and cultural achievement. In 1811, Jews achieved citizenship, in 1864 full political equality, and by 1874 Frankfurt's Jews had become so successful and assimilated that the community gradually dispersed into the city fabric. Although the Jewish community did become less visible as it assimilated in the late nineteenth century, the *Judengasse* remained Jewish space in the eyes and memories of all Frankfurters, which is to say that it served as a repository for the suppressed stories retold here. The relationship of stories and urban spaces, or history and geography, is a topic to which I will return in chapter 5. For the moment it will suffice to say that as the Jewish community prospered, it created new spaces within the city. Both the University of Frankfurt, founded in 1914, and the famed Frankfurt Institute for Social Research, founded in 1923 by Hermann Weil, owe their existence to Jewish philanthropy. The statistics are simple: in 1925, 55 percent of the Jewish population worked in banking, trade, inns, insurance, or transportation, but less than 20 percent worked in craft and industry. In that same year, 67 percent of all Frankfurt's prosperous bankers were Jews.[2]

The modern period of Jewish success and assimilation, however, came to an official end on January 30, 1933, when the Nazis seized power. The *Reichskristallnacht* pogrom of November 9 and 10, 1938, set the stage for the systematic deportation of the city's 30,000 Jewish citizens to the death camps. Approximately 100 persons survived.

This second interpretation of Frankfurt's dominant political discourse and its banking history inspired by planner Helmut Bosch is sobering but also more satisfying than the popular account found in the press and in casual conversation. Bosch argues that Frankfurt's contemporary banking industry, including its stock exchange, is not a postwar demonstration of rational invention by good Germans, but rather the direct heir of the Jewish coin exchange that emerged in the city well before 1585. In Bosch's view, Nazi anticapitalist irrationality and intolerance temporarily destroyed Frankfurt's long history of banking and cosmopolitan tolerance (Herf 1984). The simple fact is that without Nazi intervention there would have been no banks in Düsseldorf to repatriate.[3]

The point of briefly returning to this difficult history is not to reopen old wounds but rather to understand better how Frankfurt has successfully recuperated its long tradition of banking and cosmopolitan tolerance.[4] Thus, the concentration of Europe's postwar banking industry in Frankfurt was not an act of invention but an act of recuperation. This project was not one simply of reconstructing an industry dislocated by radical anti-Semitism and anticapitalism but a project of recuperating the cultural tolerance on which banking and trade depend. Put simply, money knows no color.[5] The optimization of exchange demanded by capitalism depends on the removal of those cultural barriers that would restrict cash flow to and from any quarter. In this sense, liberal capitalism is an inherently tolerant system. This observation is, however, hardly original. Karl

Marx and Friedrich Engels observed in the *Communist Manifesto*, first published in 1848, that

> [t]he bourgeoisie has through its exploitation of the world market given a cosmopolitan character to production and consumption in every country. . . . National one-sidedness and narrow-mindedness become more and more impossible, and from numerous national and local literatures, there arises a world literature." (Cited in Harvey 2000, 25)

One does not need to embrace the *Communist Manifesto* to argue that capitalism has tended to globalize local cultures and the environments they construct.

Both Marxists and capitalists commonly argue that architecture is the material embodiment of cultural politics. The patterns in which cities are constructed, the materials and technologies employed to physically build them, and how they appear to viewers is no accident of singular aesthetic whim. Rather, such constructions realize in material form the cultural aspirations and conflicts of their time. In this sense the construction and reconstruction of Frankfurt documents the struggle of Frankfurters to come to terms with global change—to tolerate not just otherness but the very idea of modernity itself. Planner Helmut Bosch alluded to this struggle in 1990, when referring to the then-proposed West End banking center, which would certainly have been very modern in its architecture and worldview. He publicly speculated whether "this amount of modernity could live wall to wall with the historic substance of the city." Bosch was, of course, acutely aware of the previous postwar reconstruction of the Römer—the city's medieval town hall—and the 1983 Disney-like simulation of the bombed-out half-timbered houses of Römerberg that sit across the square from the historic town hall. The expense that Frankfurters lavished on these architectural reconstructions reflects the determination of some citizens to reconstruct the German society that existed before the war. Some would argue that these are examples of nostalgic or even xenophobic architecture—an attempt to purify the compromised history of space.[6] (See figure 4.3.)

In his concern for the apparent modernity of the banking center plan, Bosch was equally aware of the even earlier public outrage that followed the construction of the brutalist-inspired *Technisches Rathaus* in Römerberg in the 1970s. It is reasonable to interpret these related architectural events as the public rejection of modern architecture and the social conditions presumably housed within it. Simply put, my argument here is that the reconstruction of the Römer and the recuperation of the banking center in new form are related events, or, as Bosch would have it, "two sides of the same coin." On the one side is the intolerant and xenophobic

Figure 4.3. The Reconstructed Römer
Courtesy of the author.

rejection of modernity, and on the other is the tolerant embrace of those same conditions. This dialectic pair is, of course, particularly at home in Frankfurt, which is the home of the Institute for Social Research, where dialectic critical theory emerged following World War II (Horkheimer 1947).

In this context it is important to recognize that as early as 1946 the city of Frankfurt erected a plaque at the Borneplatz to witness the destruction of its Jewish community. Nearly all German cities have, of course, made similar witness, but not as early as the citizens of Frankfurt. In 1966 a monument proposal by the Frankfurt architect J. Mayer helped to stimulate a subsequent memorial proposal by Paul Arnsberg. Arnsberg's project was finally realized in 1987 as a permanent archeological interpretation of the *Judengasse*, when the war-demolished ghetto was redeveloped for other municipal purposes. Although many Germans disagree, Helmut Bosch argues that among German cities, Frankfurt is exceptional for coming to terms with its participation in National Socialism so early. In contrast, Berlin began the process of cultural recuperation only after the capital returned to the old Reichstag after national reunification in 2000.

In sum, the return of the banking industry to Frankfurt can now be interpreted not just as the stylistic victory of modern architecture or the recuperation of liberal capitalism but as a shift in the historical discourse between the dominant values of tolerance and repressed xenophobia. The postwar era witnessed the recuperation of the rational tolerance on which capitalist societies depend to function economically. It is this conceptual link to tolerance for cultural conflict that qualifies the juxtaposition of modern banks and faux half-timbered houses as an exemplar of the style of political talk that Barber characterizes as liberal-minimalism (Barber 1984). Before reconstructing the competing story lines that framed the rebuilding of the city it would be helpful to concretely link this long history to the contemporary context of its interpretation.

Unlike in my analysis of Austin or Curitiba, Frankfurt offers an opportunity to understand public talk concerning politics, the environment, and technology through the analysis of a single case of architecture—that of the Commerzbank tower. This project was central to the city's growth pains in the 1980s and 1990s because it focused the attention of citizens, bankers, and city officials on the future of the banking industry in the city. As a result, it serves as a concrete laboratory for more conceptual planning issues. (See figure 4.4.)

4.2 THE COMMERZBANK TOWER

By the early 1980s, the Commerzbank—one of Europe's largest banks—had expanded its staff to occupy some thirty locations throughout Frankfurt. Such crowded and dispersed conditions were typical in the banking industry during that period. Planner Bosch reported that a survey by the City of Frankfurt Department of Planning in the 1980s documented a huge pent-up demand for new banking office space. This unsatisfied demand reflected, of course, historical conflict with other interests—principally those of residents in the West End district adjacent to the traditional banking area. In the 1960s the West End was home to approximately 45,000 citizens who prized the district's historic architecture, central location, and green backyards. Growing real estate pressure resulted, first, in the forced removal of many residents who rented and, second, in the 1969 creation of *Aktionsgemeinschaft Westend*, a grass-roots citizens' initiative that actively fought commercial redevelopment of the district. Nicole Weidemann, a founding member of that group, was among those who systematically reported to the city illegal demolitions or conversions of apartments into office space. Weidemann and her colleagues also documented architectural conditions that met local criteria for historic preservation as a strategy to stall redevelopment on other grounds. When the conflict over housing, now remembered by Frankfurters as the *Häuserkampf*, or housing struggle, degenerated into disruptive street violence, the radical activism of related activist groups such as *Revolutionärer Kampf*, the *Häuserrat*, and the *Putztruppe* was supported only by a few Leftist members of the SPD majority of the city

Figure 4.4. Commerzbank Tower, Sir Norman Foster & Associates. Note the random window shade pattern that reflects the ability of each office worker to control the amount of natural light and outside air that reaches his or her desk.

Courtesy of the author.

council.[7] There is, however, little doubt that such radical citizen activism made it strategically, if not tactically, impossible for growing banks to appropriate adjacent residential areas.

In the wake of the *Häuserkampf* of the 1970s, there were by the mid-1980s two competing schemes to accommodate frustrated banking interests—one supported by the conservative CDU and the other by the more liberal SPD. The CDU favored a low-rise axial development scheme that would reinforce existing transportation routes along the Mainzer Landstrasse. In hindsight, planners have recognized that such low-rise development would have effectively suburbanized the city by displacing residents to yet-undeveloped fringes of the city. The ironic rationale for this scheme was an explicit aversion to American-style high-rise density and the ecological devastation Frankfurters associated with American cities. Following the 1985 election, which was won by the CDU, Commerzbank reluctantly acquired property along the favored axis with the intent of consolidating its far-flung operations there.

The second scheme to accommodate frustrated banking interests in Frankfurt was favored by the leadership of the liberal SPD. In 1987 the SPD leader, Martin Wentz, unenthusiastically argued that any future bank construction be concentrated in the old banking quarter as a means of reducing development pressure on adjacent residential neighborhoods. This proposal stimulated intense debate within the SPD because of its implicit endorsement of "skyscrapers" (Wentz 1987). It seems that rank-and-file members of both the conservative CDU and the liberal SPD shared a dim view of an Americanized urban landscape. This political debate, however, shifted to favor high-rise development after the election of 1989 in which a coalition of the SPD and the German Green Party prevailed. The so-called red/green coalition—which I will refer to as Frankfurt's regime of sustainability—lasted from 1989 until 1995. For the purpose of this investigation, these dates are important because they were the years of Commerzbank's development. Under these circumstances, construction of the Commerzbank tower can, then, be interpreted as the *reification*, or "thing-ification," of red/green public policy.

As noted previously, Volker Hauff became the somewhat remarkable first choice for the mayor of the red/green municipal government. This choice was so interesting because Hauff—a former member of the Brundtland Commission—coauthored the first operable definition of "sustainable development" that was discussed in chapter 1 (WCED 1987). Perhaps even more unlikely was the election of Daniel Cohn-Bendit as the leading Green Party representative in the Frankfurt city council. Cohn-Bendit, a Frankfurt native, was a former student leader of the 1968 Sorbonne revolution in Paris. Under the leadership of these two progressive environmentalists, an explicit red-green agreement was reached that called for "a policy of ecological and social responsibility." It was within this hybrid political talk that the concept of the skyscraper began to take on new meaning.

Skyscrapers first became associated with social responsibility because by going up, rather than out, they avoided the destruction of the traditional neighborhoods so valued by activists. Second, skyscrapers became associated with environmental responsibility because they avoided the suburbanization required to accommodate displaced renters and the aesthetic, logistical, and social blight that comes with sprawl. Such emergent political logic mandated that "a sustainability review [be conducted] for every single [banking center] building project" (Baier 1989). Thus, third, skyscrapers became associated with environmental responsibility because the city required and bankers embraced the use of advanced, energy-efficient technologies.

Some Frankfurters to be sure—those affiliated with the romantic tradition of Goethe, the great nineteenth-century poet—argued that the skyscraper is an inherently unnatural and unsustainable architectural form. Other Frankfurters, however—those affiliated with the philosophical rationalism of the eighteenth-century philosopher Immanuel Kant—argued that the skyscraper is an inherently more efficient architectural form. These two traditions, equally parts of Germany's long internal conversation about its own character, can be understood in John Dryzek's categories as "green romanticism" and "green rationalism," respectively (Dryzek 1997).

It was into this political context that the international competition for the Commerzbank was thrown and within this context the competition was developed. In June 1990, the "Framework Plan Banking District" conducted by Novotny, Mähner, and Associates was released. The "Novotny Study" recommended a 133-meter-tall building adjacent to the historic Kaiserplatz—the current site of the Commerzbank tower. In preparation for drafting the competition brief, Commerzbank executives did apply to the city planning department to demolish several historic structures that formed the perimeter of the block purchased for development. That request, along with several other Commerzbank proposals, was heavily opposed by citizen initiatives and ultimately rejected. City planner Bosch and the news media document that intense, detailed negotiations between Commerzbank and the city lasted for several months (Göpfert 1991). At the end of this negotiation, the city and the bank agreed on a program, or description of the building, that satisfied the interests of all those sitting at the table, who included not only Commerzbank and city officials, but also social and environmental activists. The multiauthored document that resulted from this negotiation declared that the building would be a model of energy efficiency and environmental health. It would contain shops and a publicly accessible restaurant at street level, rather than an exclusive dining room on the fiftieth floor. A lecture hall would be made accessible for public programs at all hours. Parking would be reduced in favor of economic support for public transportation. And, not least important, the complex would include housing. In Bosch's view, this was to be not a bank building but a mixed-use urban development project. Nicole Weidemann, then chair of the ever-present citizen group, *Aktionsgemeinschaft Westend*, commented that the project program was "no reason to rejoice, but [it is] a compromise we can live with" (Göpfert 1991). In effect, a formula for a new kind of tall building had been constructed.

Such significant economic and programmatic concessions on the part of a developer may seem remarkable to American readers. It would, however, be quite wrong to leave the reader with the impression that citizens and a powerful municipal authority had forced a reluctant Commerzbank to invest its capital in environmental protection measures. It is certainly true that left to its own judgment the bank would not have been so socially responsible regarding making provisions for housing and public amenities. Indeed, it took "special encouragement" to obtain such concessions.[8] The employment of radically efficient environmental technology was, however, very much a part of the bank's development strategy. By 1990, Commerzbank had already incorporated environmental issues into its corporate vision. In the early 1990s, the bank created the senior post of commissioner for the environment (*Umweltbeauftragter*), who was charged with the greening of all bank operations. Environmentally inspired programs included "environmental loans" to entities unable to obtain them elsewhere, philanthropic funding of the World Wildlife Fund and the Nationalpark Bayerischer Wald, and the creation of the Commerzbank/Impulse Environmental Award for innovative enterprises.

At the beginning of the 1990s, environmental awareness was at its height in Germany. As a result, many on the Left concluded that Commerzbank was simply "green-washing" its image to exploit public sentiment. The evidence, however, suggests otherwise. An adviser to the Green Party leader Daniel Cohn-Bendit, for example, argued that "the idea to make the building green must have had its origins within the Commerzbank . . . [because city] policy didn't have that much to say [about specific environmental control techniques or social objectives] (Abele 2001). Jürgen Braune agreed that city ordinances left enough leeway to construct what could have been a very conventional building. One must conclude, then, that individual bankers, as well as the corporation, were responsible for, or at least receptive to, the environmental agenda that architect Sir Norman Foster was ultimately charged to realize as a building. The competition brief can, then, be understood as a material proposal, or recipe for conflict resolution between those affiliated with the various kinds of competing political, environmental, and technological talk that was heard around the city.

In response to the competition brief, invited firms submitted design proposals on June 3, 1991. The competition jury, however, had difficulty in reaching a unanimous decision. The jury technical adviser, Otto Frei, recommended a run-off competition between the London office of Sir Norman Foster & Partners and the German architect Christopher Ingenhoven. Commerzbank executives, however, refused to cooperate—as did Foster—and amid some controversy, Foster was selected by mandate of the bank. Although Ingenhoven's proposal may have been the more technically advanced, beginning on June 29 the popular press described Foster's "winning" design as an "ecological high-rise" (Frei 1991) and an "eco-skyscraper." Martin Kohlhaussen, the spokesman of the Commerzbank board of directors, confirmed to the world that "[i]t is all about ecology" (Lattka 1991).

It would be, I am afraid, overly tedious to many readers to enumerate in detail the environmentally inspired technologies that were employed in the building since these are documented elsewhere (Davies 1997; Pepchinski 1998). Nevertheless, the most significant deserve mention: the so-called double-leaf facades that employ solar energy to warm an air-blanket around the building; district heating distributed by the city; absorption chilling in lieu of compression chilling typically used in the United States; hydronically cooled radiant ceilings rather than air-conditioning; lighting integrated with ventilation to absorb the heat of occupancy; passive nighttime thermal mass charging that employs natural nocturnal temperature swings; and an energy management system that operates windows. These are exotic technological choices by American and Brazilian, if not German, standards.

Perhaps the most innovative aspect of the building's design is the naturally ventilated atrium and system of "sky-gardens" developed by Foster (see figure 4.5). That all workstations are no further than nine meters from an operable window, thus guaranteeing employees access to natural light and ventilation, is a feature that has captured the enthusiasm of the public. Although workspaces without natural light or natural ventilation are typical in American office buildings, they are actually illegal in Germany—a result of legislation initiated by strong office-worker unions. In many respects the constructed building does not just satisfy code requirements, but actually exceeds them. Architecture critics have argued that the design team's manner of satisfying German environmental codes amounted to nothing less than the redescription of the skyscraper as a building type. Such an enthusiastic reception of the building by architects and critics is an indicator that the integration of environmental, techno-

4.5. Commerzbank Interior Atrium
Courtesy of the author.

logical, and programmatic innovations resulted in aesthetic opportunities that were fully appreciated and exploited by the designers.[9]

Dr. Arnheim Müller, director of development for Commerzbank, has documented that in 2002 the energy performance of his building was 30 percent more efficient than buildings of the same type built in the same period—the Messeturm by Helmut Jahn, for example, was just completed when the Commerzbank was in early design stages. Perhaps more significant, however, is the observation made by city planner Bosch. He argues that the public talk that accompanied the development of the Commerzbank building stimulated a new era of energy-efficient architecture in Frankfurt. Although Bosch recognizes that the Commerzbank is a far more satisfying example of tectonic clarity than the projects that immediately followed it, he claims that more recent projects—the Maintower, for example—are actually more energy efficient than the Commerzbank tower. Just as the banks of Frankfurt, like anywhere else, vie to build the highest tower and the most recognizable crown, unlike elsewhere they also vie to publish the lowest energy consumption rates in an effort to gain favor in the streets and at city hall. In this limited sense, it is fair to argue that the Commerzbank tower has materialized the public talk of its time.

4.3 FRANKFURT'S DOMINANT STORY LINE —PROGRESSIVE CAPITALISM

At this point in reconstructing Frankfurt's story it will help to relate its history to the categories introduced at the beginning of the chapter. As in Austin and Curitiba, I will argue that the end of the twentieth century in Frankfurt witnessed the confrontation of two competing story lines—the dominant story line, which I refer to as *progressive capitalism*, and the counter-story line, which is best described as *red/green*. The dominant story line, progressive capitalism, can be said to be a Western European phenomenon and be characterized as a brand of twenty-first-century capitalism that managed to internalize many of the critiques that socialism threw its way in the nineteenth and twentieth centuries. Thus, life in Frankfurt is framed by a story line that is regional in scope. In contrast, Frankfurt's red/green counter-story line was locally constructed and novel in the Europe of that era. Although the red/green government lost the municipal election of 1995, similar coalitions have emerged in other cities by knitting together local political, environmental, and technological talk. As in chapters 2 and 3, I will reconstruct the dominant and counter-story lines one at a time, beginning with the dominant version.

Frankfurt's Dominant Political Talk

I argued earlier, on the basis of the city's long history, that the dominant political talk in Frankfurt is what Barber refers to as "liberal-minimalism" (Barber 1984). In more common language we can describe Frankfurt as a "pluralist" democracy in which the inevitable conflicts that occur between groups and individual citizens are resolved through bargaining—a cultural norm, or habit, that is consistent with the city's role as a regional, national, and international marketplace. The integrity of bargaining is governed by an implicit social contract that is based on tolerance. The contract constructed by citizens is to agree that "so long as you play by the rules, you're welcome to play the game." As in Austin, Frankfurters value highly their freedom to pursue private interests in the context of the market. Unlike Austin, however, there are more complex rules for engagement in the public space of the market that operate to restrict some individual rights. The rules—enforced by the "invisible hand" of liberal economic theory—claim to ensure relatively equal access to the market by all economic actors and thus benefit society as a whole. It is the same social contract that also claims to optimize exchange among the many rather than optimize accumulation by the few. On the basis of the previous abbreviated, yet still "thick," history of banking and tolerance, it seems reasonable to conclude that the dominant political discourse of Frankfurt is indeed liberal-minimalism. Arnheim Müller provides two arguments that serve to correlate this claim to the case of the Commerzbank tower.

First, Müller claims that building the tower was a significant move toward realizing the compact city as a setting for the "new economy"—a term then used to describe the increasing flexibility of digital capital. Müller, like many others in that era, associated digital media with equity and tolerance because it was understood to provide anonymity and thus access to the market by marginal economic actors (meaning women, people of color, and nonnative speakers of German). Economic development can, according to this logic, be seen as advancing the individual interests of those who might otherwise be excluded. Second, Müller argued that the open-space interior planning of the bank's offices, which was highly unusual in Germany at the time, was evidence that the bank sought equity for all employees and was working toward creating nonhierarchical space. This logic is to suggest that the "invisible hand" of liberal economic theory tends to flatten rather than heighten corporate hierarchies. Taken together, these two claims by the manager of the tower project are indeed consistent with the logic of liberal-minimalism.

Frankfurt's Dominant Environmental Talk

The dominant environmental talk to be found in Frankfurt is the "weak" version of what Dryzek (1997) terms "ecological modernization." In Dryzek's taxonomy this kind of talk is closely related to that of "sustainable development" as it is articulated in *Our Common Future* (1987) and generally argues in favor of reforming modern institutions to work within newly discovered ecological limits (Mol 2003). In the view of those who support ecological modernization, emphasis should be placed on simple resource sufficiency, not abstract or misguided spiritual agendas.

It is particularly interesting that the Commerzbank executive, Arnheim Müller, who is a good spokesman for ecological modernization, agrees with the romantic critics of his project—that there can be no such thing as a "sustainable high-rise." His assessment, however, seems intended to discredit the romantic view as archaic and impractical rather than suggest a common cause between the two interpretations. In the assessment of those who subscribe to talk about ecological modernization, the rationale behind the project is not environmental protection so much as productive efficiency realized through reduced energy consumption and increased worker productivity—a kind of "clean and green capitalism" (Dryzek 1997, 149). Although there has been no attempt at postoccupancy analysis to document increased worker efficiency, Commerzbank executives have discussed such a project with faculty at Carnegie-Mellon University in Pittsburgh. Müller is indeed keenly interested in the possibility of rationalizing the bank's investment in environmental quality as a form of indirect—and thus nontaxable—worker compensation.

Müller and his banking colleagues recognize that the technologies they helped to create participate in complex natural systems by using nature as a sink. They also understand that to take advantage of nature's capacity to absorb wastes requires complex partnerships between government, business, science, and environmentalists that are cooperative, not antagonistic. It is such cooperation, in their view, that enables progress toward wealth production within defined ecological limits.

Frankfurt's Dominant Technological Talk

The technological talk that I have reconstructed from my investigation of Frankfurt is a *soft* version of technological determinism. The *hard* version of this kind of talk is to argue that technologies emerge from some predetermined logic that is present in artifacts themselves and that society is shaped by whatever is next on the list. Hard technological determinists tend to see history as a story whose ending was inevitable from the beginning because they project backwards the logic of technological systems and see them as causes for development rather than by-products. One very unfortunate consequence of this logic is that citizens become unable to imagine alternative futures.

Arnheim Müller can again serve as an articulate spokesman for the dominant talk, soft technological determinism. There is no doubt in Müller's mind that technology will solve our environmental problems. He is, however, not naïve about how technological choices are made. Optimal choices, far from being the next thing on some predetermined list, are designed, he believes, by teams of experienced experts. He is generous with his assessment of how the architects (Foster & Partners) and engineers (Ove Arup & Partners) performed for the Commerzbank and refers to them collectively as "the best out there today." He quickly follows up, however, by saying that "Foster is not a super-architect. You must tell him, more or less, what to do." This bit of bravado is mediated by Müller's detailed description of how he, the architects, the engineers, city planners, and a very long list of suppliers and manufacturers negotiated in excruciating detail over a period of two years how the building would actually be built. Because of the stature of the project, no fewer than 120 engineering studies were conducted, some by the city, some by the design team, and others by suppliers to determine whether the intentions of the competition brief written by the red/green coalition could be met at a cost that the bank could afford. Müller argues that, in the end, the completed project is so exemplary only because the design team functioned so cooperatively. A social scientist might agree with Müller's assessment but would place far greater emphasis on how the economic and political interests of the various groups guided building design rather than on building design as an autonomous activity.

In general we can say that this team of soft technological determinists is technophilic, or that they are lovers of technology, but they also recognize the environmental and social limits to invention. They see themselves as being responsible for moderate progress toward ecological improvement, but only when market and political conditions allow for advanced levels of collaboration. Although these experts are drawn toward tools that are high-tech, part of their value system is to employ only those tools that can be produced and used profitably.

The progressive capitalists of Frankfurt, who dominated the scene before and after the emergence of the red/green regime of sustainability can, then, be fairly characterized as socially tolerant, instrumental managers of the environment, and firm believers in the power of technology to solve our problems. These characteristics are summarized in table 4.1.

4.4 FRANKFURT'S COUNTER-STORY LINE—RED/GREEN

The red/green coalition of democratic socialists and environmentalists ushered into power in Frankfurt's municipal election in 1989 served until 1995 and in that period constructed what can be understood as the city's regime of sustainability. Where progressive capitalists had been tolerant to a fault, the red/greens sought social transformation. Where progressive capitalists were instrumental managers of natural resources, red/greens sought to alter the categories that distinguished between human and nonhuman well-being. And where progressive capitalists embraced advanced technology as the solution to the city's problems, red/greens sought to invent technologies that would enable a particular way of life. Construction of the Commerzbank tower is one of several examples that might document this difference between regimes.

Table 4.1. The Dominant Story Line of Frankfurt—Progressive Capitalism

Kinds of talk	As told by: Herbert Abele, Helmut Bosch, Jürgen Braune, Norman Foster, J. Koolhaas, Arnheim Muëller, and Uwe Neinstedt
Dominant Political Talk	*Liberal Minimalism*
Generic response to conflict	Toleration— Free individuals resolve conflict through bargaining and negotiation
Attitude toward certainty	Skeptical— In pursuit of optimization
What is valued	Freedom— To pursue individual happiness in the market
Conceptualization of space	Active— Individuals in motion naturally come into conflict
Dominant Environmental Talk	*Ecological Modernization*
Basic entities recognized or constructed	Complex systems and nature (as a sink) are recognized as entities within a liberal capitalist economy mediated by the state
Assumptions about natural relationships	Partnerships between government, business, science, and environmentalists are cooperative; society subordinates nature, but environmental protection and economic development are complementary
Agents and their motivation	Progress requires complex partnerships, and these are motivated by individual benefit and the public good
Key metaphors and rhetorical devices	*Oikos*—a tidy house Progress— Reassurance that we can have both wealth and nature
Dominant Technological Talk	*Soft Determinism*
Attitude toward technology	Technology can solve our problems, but there are economic, environmental and social limits to what can be done
Attitude toward history	Progress is possible
Assumptions about the origins of technology	Technologies are developed by teams of experts in response to market conditions
Types of tools employed	Those that can be produced and used profitably

Frankfurt's Political Counter-Talk

In Barber's lexicon, Frankfurt's red/green coalition is as close as we will find in this study to his proposal for "strong democracy." Those who generally subscribe to the political counter-talk of strong democracy and who are knowledgeable about the Commerzbank tower, include Barbara Holzhausen (Green Party), Käthe Schröder (architecture critic), and Nicole Weidemann (SPD activist). In contrast to Arnheim Müller, these observers tend to view the culture housed in and enabled by the Commerzbank tower not as equitable and tolerant but as the exclusive domain of elite corporate workers. The process of its construction, they argue, is a case in point: although the federal law mandates a two-step development process that includes public participation, the level of public participation was experienced by these actors as perfunctory rather than substantive. Even within the Commerzbank hierarchy this view holds true. Müller, for example, confirmed that bank employees had virtually no input into the project scope, program, or design. Such employee input would have, he argued, slowed the development process unacceptably. An accounting of the claims made by Müller is tallied differently by the assessment of skeptics. They argue that, yes, the bank has a restaurant open to the public, but it is architecturally isolated from the street because it was raised to the second level and thus effectively private. On Saturday, August 4, 2001, for example, there were three tourists in the café sipping coffee at 1:00 p.m. because there was no food being served on a day when few bank employees were in the building. At the same time, adjacent cafes on the Kaiserplatz were overflowing. Yes, there is housing included in the project, but 4,500 square meters, or sixty to seventy units, is a drop in the bucket compared to the number of units that were displaced by related commercial development in the neighborhood.[10] And, yes, there is an auditorium available for public use, but almost no one knows about its existence or how to book the facility. And, yes, Commerzbank workers have the benefit of relatively equal and architecturally splendid working conditions with natural ventilation and natural light, but these are generally elite executives who understand this environment to be a form of exclusive compensation that is not available to lower-level Commerzbank workers in other locales. In fact, more recent Commerzbank office space constructed in Frankfurt for lower-level workers does not meet the same environmental and social standards. In this sense, space within the Commerzbank corporate complex is very hierarchical rather than equitable. In the eyes of strong democrats, the program of the tower was certainly transformed through public talk, but not nearly enough to satisfy their aspirations.

Those who subscribe to the political counter-talk of strong democracy in Frankfurt operate with very different assumptions than do those who subscribe to liberal-minimalist values—they see in conflict an opportunity to transform conditions rather than an opportunity to demonstrate tolerance. The experience of the *Häuserkampf* is a prime example. Had neighborhood activists only tolerated the conflict between residents and bankers the latter group would surely have prevailed in the struggle over the West End's real estate. Tolerance is surely an admirable characteristic, but in itself it is powerless to transform conditions (Rorty 1998). The vocal intervention of activists, however, transformed the architectural program of the tower from that of a typical bank building to a mixed-use urban structure that included housing, retail, public facilities, and substantial historic preservation of existing buildings on the site. It is also fair to say that their intervention required the use of previously unknown or untested technologies. Their intervention, then, was an example of transformative politics, not toleration of conflict.

Affiliates of the red/green regime of sustainability never became rigidly certain of the goal they pursued through twenty years of confrontation and public talking. Rather than ideological, their goals were—oddly like those of the Lerner regime in Curitiba—ad hoc and limited in scope. Although the drafting of the architectural competition brief appears pivotal in my reconstruction of this story, there are many other events that are equally significant. These local victories were part of a loose coalition of kindred spirits rather than an ideologically coherent movement set on a utopian future. What was valued in this long process was the participation of citizens, construction unions, bankers, planners, environmentalists, and radicals of all stripes. This kind of inclusiveness requires that urban space be conceptualized not just as active but also as a continuous mix of natural and cultural flows. It was this strongly democratic story line that defined the terms and aspirations of the architectural brief that, in turn, encouraged designers to interpret social and environmental conditions in dramatically novel ways.

Frankfurt's Environmental Counter-Talk

In chapter 1, I argued that there are always multiple kinds of talk, even within a single conversation. This is especially true when trying to describe the relation between humans and nature. The geometry of Frankfurt's environmental story line is unlike those of Austin and Curitiba because there are two equally viable kinds of counter-talk that vie for the allegiance of supporters in the struggle against the dominant environmental talk, ecological modernization.

As I argued earlier, both "green romanticism" and "green rationalism" are long-standing traditions of interpretation in Germany and are roughly congruent with the two factions of the German Green Party, the *Realos* (or realists) and the *Fundis* (or fundamentalists). I will reconstruct the values of each group beginning with green romanticism, because it provides a background well understood by most, if not all Germans.

The green romantics in this case include Uli Baier (Green Party representative to the City Council), the Green Party's Barbara Holzhausen, and the architecture critic Käthe Schröder. These *Fundis* argue that public talk in Germany regarding environmental protection was at its peak in the years leading up to the Novotny Report (1990). The subsequent period in which the Commerzbank tower was in its formal planning stage (1990–1994) was, in their romantic view, strongly influenced by the bank's desire to court public approval through a public relations campaign designed to represent the bank as a good environmental citizen—in other words, to instrumentally paint itself "green." It is these same observers who tend to argue that a "sustainable high-rise" is an oxymoron by definition—meaning that any tall building is disproportionately resource consumptive and must result in an unhealthy microclimate. By extension, this group tends to see cities themselves as inherently anti-ecological. It should come as no surprise that the romantics associate sustainable, or green, architecture with more rural or naturalistic conditions. The ideological purity of their position requires them to reject the seemingly pale green efforts of the bankers as simple opportunism.

Many readers may recall that the great German romantic poet mentioned earlier, Johann Wolfgang von Goëthe (1749–1832), was a Frankfurter. I recall this fact not to suggest that all Frankfurters are green romantics but to argue that the romantic critique of modernism is an important thread in German history. In this larger context, their critical interpretation of Commerzbank's environmental performance can be understood as a general rejection of Enlightenment values. Observers on both the inside and outside of development decisions view the project as architecturally accomplished but ultimately an economically inspired experiment in energy efficiency and long-term cost reduction, not an experiment in environmental preservation inspired by a poetic or spiritual connection to nature. Readers will also recall that green romanticism was a powerful counter-discourse in Austin, so I will not further analyze the values that lie behind it. It will suffice to say that green romantics tend to construct an ideological landscape in which they are the deepest shade of green along a continuum of successively lighter values (Naess 1995).

The second ecological counter-talk in Frankfurt is articulated by Jürgen Braune and Helmut Bosche. Dryzek (1997) would describe their set of values as "green rationalism," which is roughly synonymous with the position taken by the German Green Party faction known as *Realos*. Where green romantics can be characterized as rejecting politics and

Enlightenment values, green rationalists can be characterized as embracing them—*Realos* are both environmentalists and humanists.

For these rational actors in our story it is no longer appropriate to demonize corporations as evil capitalist institutions or their managers as distant profiteers. The green rationalists argue that there is a clear "red line" of cultural continuity from the political activism of 1968 into the twenty-first century. Although the radical ideals of that earlier period have certainly mellowed with age, the green rationalists argue that the same University of Frankfurt students who occupied the barricades in 1968 now occupy both the boardrooms of the Commerzbank and the halls of political power. Daniel Cohn-Bendit and Joschka Fischer are prime examples. Cohn-Bendit has had a distinguished career as a representative of the Green Party to the Frankfurt City Council and, subsequently, a German Representative of the Green Party to the European Parliament. In a similar ascent, Fischer became German foreign minister in 1998. Both were active participants in the public talk that preceded the Commerzbank tower project. According to Barbara Holzhausen, Cohn-Bendit publicly argued against any building greater than three stories in height in the early 1980s, but by 1994 he became a principal champion of increasing the height of the Commerzbank tower to 185 meters. Müller confirmed the positive role that Cohn-Bendit played as a Green Party member of the City Council in permitting greater building height in exchange for employing more environmentally friendly technologies and providing more public amenities.

Braune and Bosch are quick to distinguish the confrontational politics of the 1960s and 1970s from the more conversational, or cooperative, politics of the twenty-first century—the latter is, in their view, far more productive. Braune and Bosch do not naïvely accept Commerzbank's environmental intentions as disinterested—far from it: they appreciated (and enjoyed, it seems) the political nature of the lengthy negotiations in which environmental and social objectives have been realized by granting the bank significant economic concessions. In the end, however, these sophisticated political actors are willing to accept as genuine the environmental intentions of particular corporate managers.

Unlike Bosch (the planner), Braune (the Greenpeace activists) is skeptical of the power of government to do much at all. He is, however, very optimistic about the prospects for nongovernmental organizations to negotiate explicit social equity issues directly with corporations and unions. The Greenpeace Clean Construction initiative is a case in point. In the program, professional consultants retained by Greenpeace assist union construction workers affiliated with the IG Bau—Europe's largest construction union—inspect construction sites and thus reduce environmental and health hazards while increasing energy efficiency.[11] This is certainly an example of red/green values in which environmental health and worker health are considered an indivisible whole—a productive conflation not like to happen in either Austin or Curitiba.

In general we can say that green rationalists recognize complex ecological systems and their limits in relation to social, economic, and political structures. They differ from other participants in the Commerzbank project in that they insist on equity among all humans. Progress toward distributive equity, they argue, is a complex process in which the relations between human and natural systems must be rationally considered.

Frankfurt's Technological Counter-Talk

Earlier, I characterized the dominant technological talk in Frankfurt as a soft version of technological determinism. The opposing kind of talk, which can be equally extreme, might be called social determinism, but I prefer to use the term *technological voluntarism*, because it suggests that humans can voluntarily select the technologies they want. In its most extreme form, technological voluntarists imagine that history presents no obstacle to our choices. In the case of Frankfurt and the Commerzbank tower, the determinist and voluntarist positions are softened by both the political and the environmental story lines discussed previously.

Counter to Frankfurt's soft determinists are its soft voluntarists. The distinction to be made between these two ways of interpreting the world of technological change is really a matter of emphasis rather than foundations. Whereas determinists focus on the influence that technologies have on society, voluntarists focus on the influence that society might have on technologies. In the case of the Commerzbank tower, planner Helmut Bosch best articulates the values of this group that includes Green Peace's Jürgen Braune and activist Nicole Wiedemann.

Bosch is quick to point out that there is no political distinction between choosing technologies at the scale of the city or at the scale of the building. Unlike in the United States or Brazil, nearly all buildings in Frankfurt's core participate in districtwide infrastructures, including heating and cooling systems, and are often energy producers themselves. This condition means that private buildings are a part of the public infrastructure and that the categories of public and private are blurred by the social history of technology. As a result, it is impossible to know exactly where the line between public and private occurs—it is a matter of bargaining, not of legal principle as in Austin. This mechanical condition serves to both reflect and realize the degree to which the public sector is engaged in making technological choices in private buildings. Germans in general and Frankfurters in particular do not recognize the rights of individuals to make whatever technological choice meets their immediate needs. Of course, this is not the situation in Austin or Curitiba, but in Frankfurt there is dramatically greater focus on the role that public-sector planners have in making technological choices that affect public space.

In the case of the Commerzbank, I have already mentioned that more than 120 engineering tests were conducted to assess the public consequences of private construction. As in other large Frankfurt projects, many of these tests were conducted by city staff or their consultants so as to build an ever-growing body of knowledge that will inform future urban decisions. Bosch makes it very clear that the city technical staff has considerable authority to negotiate the environmental and program characteristics of any given project. Their authority to negotiate the functioning of systems, however, is based on performance criteria that are determined by the public political process. Another way to say this is that urban codes are in a constant state of development and are performance-based rather than prescriptive. This means that the city legislates the desired outcome of building, not particular construction techniques as is the case in Austin and most American cities.

Frankfurt, then, has developed a planning culture that is not unique in the context of the EU but very different from that of the other two cities being investigated. The history of planning in Frankfurt is old and distinguished, particularly in the Weimar era (1919–1933) in which architect-planner Ernst May (1886–1970) and his colleagues established a social rationale for determining urban form. Building on May's insights into urbanism, urban planning in Frankfurt is now practiced as the building of flexible codes that are outcome oriented rather than prescriptive. Such codes reflect and reinforce the political goals articulated by citizens and their representatives in the city council. The role of planners in Frankfurt is, then, very different from that in Austin, where planners struggle to keep up with the consequences of market-controlled development, or Curitiba, where planners determine market opportunities. The characteristics of the dominant and counter-story lines of Frankfurt are reconstructed in table 4.2.

Table 4.2 The Counter-Story Line of Frankfurt—Red/Green

Kinds of counter-talk	As told by: Uli Baier, Helmut Bosch, Jürgen Braune, Danny Cohn-Bendit, Joschka Fischer, Barbara Holzhausen, Käthe Schröder, and Nicole Weidemann

Political Counter-Talk	Strong Democracy
Generic response to conflict	Transformation— Public talk resolves conflict by transforming awareness
Attitude toward certainty	Ontological— In pursuit of process
What is valued	Participation— Individuals become free by governing themselves at least part of the time
Conceptualization of space	Continuous— Individuals exist within a continuum of flows

Environmental Counter-Talk Number 1	*Green Romanticism*
Basic entities recognized or constructed	Global limits, inner nature, unnatural practices, and ideas
Assumptions about natural relationships	Wholeness— Relations between humans and nature should be "natural"
Agents and their motivation	Humans and nonhumans— Both groups are subjects
Key metaphors and rhetorical devices	Organicism and passion— Appeals to the emotions and intuition

Environmental Counter-Talk Number 2	*Green Rationalism*
Basic entities recognized or constructed	Global limits and nature as a complex ecosystem are recognized in relation to social, economic, and political structures
Assumptions about natural relationships	Equity should exist among all people; there are complex interconnections between humans and nature
Agents and their motivation	Many individuals and collective actors are at play with complex motives
Key metaphors and rhetorical devices	Organic— With appeals to the rationality of social structures. Progress

Technological Counter-Talk	*Soft Voluntarism*
Attitude toward technology	Choosing a technology is choosing a way of life; we can have the technologies we want, within economic and political limits
Attitude toward history	Progress is possible
Assumptions about the origins of technology	Technologies are socially constructed
Types of tools employed	Those that will enable a desired way of living within sustainable environmental costs

4.5 SUMMARY

The conflict between progressive capitalism and the red/green story line in Frankfurt is substantial but not nearly as severe as that between the dominant and counter-story lines in Austin and Curitiba. Through hermeneutic lenses it seems reasonable to interpret the relative closeness of world views in Frankfurt as an already narrowed horizon of possibilities. For progressive capitalists, the future promises that the once-crude institutions of modernity will be socially reformed and made more efficient by technology. Banks, for example, will become more socially and environmentally reflective, yet remain responsive to the laws of supply and demand. In contrast, the future promised by the red/green story line promises reform of the life-world through citizen participation in redescribing the technologies we need to live sustainably. The distinction between "redescribing" a technology (the goal of red/greens) and "inventing" one (the goal of progressive capitalists and an explicit claim made by Foster & Partners) has to do with ones view toward history. The claim to inventing a thing is historical—it presumes a new beginning. In contrast, the redescription of a thing acknowledges its prior existence before it can be redescribed (Rorty 1982). The distinction suggests that an act of redescription is a narrative of political, environmental, and technological change in which designers play an active role in materializing possible urban futures.

The structure of Frankfurt's two story lines and seven kinds of talk are summarized in tables 4.1 and 4.2. As in chapters 2 and 3, the tables provide a taxonomic view of competing possibilities for the city. The danger of taxonomies, however, is that they can restrict rather than expand possibilities for meaning, so as in the previous chapters, I will propose a few of the lessons that might be learned from Frankfurt's story. There are four: the first concerns coalition building, the second conflict resolution, the third economic determinism, and the fourth redescription through the use of technological codes.

Coalition Building

The red/greens of Frankfurt are an excellent example of coalition building in its most productive form. The reds derive their worldview from Germany's Marxist tradition and are ideologically committed to the modern project of human liberation. In contrast, the greens are already a coalition of *Fundis* (environmental fundamentalists) and *Realos* (environmental rationalists) that derive their respective worldviews from Germany's romantic and Kantian traditions. In the context of Frankfurt's long public conversation concerning urban form, all three of these groups had become well practiced in collaborative action, if not in constructing common ideals. This is to say that public talk has become for these activists a comfortable recreation in which short-term strategic alliance might be made without causing the gut-wrenching anxiety experienced by ideological purists who can talk only in their own frames of reference. Common actions, even those derived from opposing ideals, are understood by them as small experiments that lead to new opportunities, new projects, and new coalitions. In the process, flexible minds (like that of Danny Cohn-Bendit) are free to learn from their experience and the evidence mustered by colleagues from other traditions. Coalition building, at least in this sense, does not reflect unprincipled opportunism or flip-flopism, as some purists have suggested, but a healthy understanding that interpretations change with changing conditions.

Conflict Resolution

Frankfurt's regime of sustainability was brought into existence by a coalition of the left-of-center SPD and the German Green Party. Some might argue that there are many possible coalitions that might have overcome the traditional dominance of business and real estate interests in municipal governance. In chapter 5, I argue against this thesis and in favor of the red/green coalition as a necessary if insufficient condition for sustainable development to occur. But the case of Frankfurt also gives rise to a caution—that of cooptation or mere compromise in the process of resolving conflicts.

In his study of "technology-oriented social movements," David Hess defines such social phenomena as "organic food," "alternative medicine," or "New Urbanism" as social movements sharing two fundamental characteristics: first, opposition to a specific existing technology and the development of alternative technologies that become the basis of industrial reform and second, incorporation and transformation of the alternative technologies by established industries (Hess 2002). Hess finds a mixture of success and cooptation no matter which side of the conflict you support. Grassroots organizations like *Aktionsgemeinschaft Westend* were only mildly pleased with the outcome of the Commerzbank project because their insistence on no net loss of housing in the West End was not satisfied, but neither was it ignored. The same lackluster response to partial victory was heard from bankers as well as romantic environmentalists. The technologies that finally showed up at the Kaiserplatz were neither perfectly efficient nor without entropic consequences. Rather, the incorporation and transformation of alternative technologies into dominant rather than marginal story lines necessarily alters their form in the process of altering the story lines themselves. My point here is not that compromise between apparent enemies is inevitable but that coalitions between unlikely partners fosters new stories.

One cannot help but be impressed with the degree of conflict resolution achieved in the process of developing Frankfurt's banking landscape. Although the articulation of differences in 1980—the height of the housing conflict—was distinct and powerful, by 2001 respondents from each interest group voiced only minor dissatisfactions. This may well be a sign that most interests were satisfied and that no principal interest was vanquished. It may also be a sign that weaker interests were suppressed. Of course, there were many less visible interests, such as those of the Jewish community, bank employees, and dislocated renters who were never invited to participate in the discussion in the first place.

The representational function of the architecture, however, particularly of the Commerzbank tower, played no small role in the resolution of old conflicts. I found that even those citizens who were initially hostile to the concept of skyscrapers in Frankfurt have grown to identify their own interests with them. The new city skyline has become a collective source of pride because it distinguishes, through architectural form, the unique role of the city in the global economy. It is this representational, or aesthetic, function of architecture that may approach, if not entirely satisfy, Barber's desire to "transform" social consciousness and get beyond the atomized quality of liberal-minimalist society. That Frankfurters articulate a new sense of collective identity materialized in tall buildings is a powerful critique of the Lockean version of liberal capitalism in which the whole of society is never more than the sum of its individual parts. Beauty, it seems, has social consequences.

It is important to distinguish how architectural representation in Frankfurt operates differently from that in Curitiba. Banks are, of course, acutely aware of architecture as a medium of advertising and product differentiation in a competitive market, and those in Frankfurt are no less self-conscious than elsewhere. The difference is that the banks of Frankfurt can produce data that document their reduced burden on the ecosystem and thus on their fellow citizens. The public buildings of Curitiba cannot because they employ conventional, even if expressive, technologies. As a result, the symbols employed by the architects of Frankfurt's banking landscape have come to represent environmental values that citizens endorse. In contrast, the symbols employed by the architects of Curitiba's urban landscape have come to represent the *regime,* not the *practice,* of sustainability. In both cases the local meaning of architectural symbols is socially constructed, meaning that there is no deep significance that links any particular sign to its referent. The distinction is that the meaning of the architectural signs deployed in Frankfurt—the triangular plan of the building, for example—was constructed through a twenty-year public conversation, not found whole in a universal meaning ascribed to triangles. My point is that Frankfurters consciously resolved conflict through the transformation of the technological systems embodied in skyscrapers. By talking about how they would like to live, citizens catalyzed new possibilities in technology and architecture. Fortunately, these possibilities were entrusted to architects and engineers capable of seeing and realizing them.

Economic Determinism

In all the data collected in Frankfurt, however, the counter-talk of green romantics and green rationalists rings most true—there is an underlying rationale in the remaking of the city that I can describe only as *economic determinism*. This term suggests that any concern for beauty, environmental preservation, or social equity is always already prefigured by what Helmut Bosch described as "economic calculus." Benjamin Barber associates this phenomenon with liberal-minimalist regimes where, in his language, "economics precedes politics" (Barber 1984, 252). He means by this comment that environmental or social equity arguments are tolerated but always trumped by economic arguments. Barber's argument for strong democracy, of course, requires the reverse, that politics precede economics. Economist James Galbraith makes a similar but more expansive point by arguing for the autonomy of different realms of life (Galbraith 2001). The problem for Galbraith is that in liberal-minimalist regimes economic values have come to dominate not just the environment and the calculation of social equity but also the intellectual, artistic, and even religious realms of public life. Such domination is clearly *hegemonic* in that it has come to be understood as natural even by those who suffer most.

The arguments against economic determinism made by Barber and Galbraith are, however, not exactly new. Marx and Engels made a similar critique of American capitalism in the nineteenth century. Although their critique may overstate the conditions found in contemporary Frankfurt, the globalization of American-style capitalism justifies their claim that liberalism in general

> has left remaining no other nexus between man and man than naked self-interest, than callous "cash payment." It has drowned the most heavenly ecstasies of religious fervor, of chivalrous enthusiasm, of philistine sentimentalism, in the icy water of egotistical calculation. It has resolved personal worth into exchange value, and in place of the numberless indefeasible chartered freedoms, has set up that single unconscionable freedom—Free Trade. . . . The bourgeoisie has stripped of its halo every occupation hitherto honored and looked up to with revered awe. It has converted the physician, the lawyer, the priest, the poet, the man of science, into its paid wage labourers." (Cited in Harvey 2000, p. 22)

Following this historic objection of the Left requires that any critique of Frankfurt, and of the model of sustainable development offered up by liberal-minimalists there, must be based on the inherent imbalance imposed on the triangulated model of sustainability by the economically deterministic logic of liberal capitalism. This is to say that equating economic development with either ecological integrity or social equity is a fundamentally unsound calculation because these are not commensurable values. The argument is to reject the notion that either life itself or justice has a value that can be expressed in purely economic terms. This almost banal finding will come as no surprise to most readers (excluding neoclassical economists), yet its implications are no less significant. If the interests of economic development always prefigure the value found in the remaining two points of the triangle one can never end up in the center. This finding suggests that the sustainable development model itself is flawed—a consideration that we will return to in chapter 6.

Political conditions in Germany, however, provide some reason for optimism. Planner Bosch argues that the German constitution and the federal planning law have already at their disposal mechanisms intended to balance the demands of commonweal and private property. The concept of *Abwägung*, or "weighing up," gives planners the authority to tip the scales, if you will, in favor of those interests that are less forcefully represented but considered valuable to community life. The only question, then, is if cities choose to employ the balancing tools that are constitutionally available. In common practice architects, planners, and citizens can only hope that the emerging discipline of environmental economics, or consumer demand for green products, will adequately rationalize any investment in environmental or social health. Although progressive capitalism has clearly succeeded in institutionalizing tolerance as a condition of modern life, it has done so—as Galbraith has argued—at the cost of every other sphere of life. The hope of activists such as Bosch is that *Abwägung* will guide future public talk away from the logic and practice of economic determinism that has dominated the first fifty years of development in the new German nation.

Code Making

One can describe the development of the Commerzbank tower as what Andrew Feenberg would characterize as a case of "subversive rationalization" (Feenberg 1995), by which he means the very idea of what a tall building might be was subversively redescribed by twenty years of public talk by Frankfurters concerning the desired qualities of urban life. The American model of a skyscraper was not simply appropriated by Frankfurters for economic reasons and then redecorated with new architectural signs to render it German, or green-washed with a few solar collectors to render it sustainable. Rather, this building type was reimagined through the social construction of new technical codes (Feenberg 2002). The fact that virtually every workstation in the fifty-story tower has access to natural ventilation and natural light is unimaginable to American architects and engineers. It is this self-conscious process of redescribing the city and its building types through a public code-making process that subverts both technological and economic determinism. In other words, Frankfurters were able to imagine an alternative future and were sophisticated enough to understand that it could be realized not by making the building *look* different but by establishing new technological norms or building habits that would make it *live* different. As Feenberg (1995, 10) argues, "Technology is not merely the servant of some predefined social purpose; it is an environment within which a way of life is elaborated." The Commerzbank tower competition brief can in this context be understood as one successful attempt by the red/green regime of sustainability to elaborate a sustainable way of urban life.

The lessons learned in Frankfurt related to coalition building, conflict resolution, economic determinism, and code making are particularly helpful in understanding the route taken by that city toward sustainable urban development. I have made comparable observations derived from the study of Austin and Curitiba. Before it will be possible to synthesize these lessons into abductive proposals for action, it is necessary to employ the spatial tools of geography to confirm what I think has been learned from history and social science.

CHAPTER REFERENCES

Abele, H. (2001). Telephone interview, 12 November.

Baier, U. (1989). "Six new high-rises." In *Frankfurter Runschau,* 20 April.

Barber, B. (1984). *Strong democracy: Participatory politics for a new age*. Berkeley: University of California Press.

Davies, C. and I. Lambert. (1997). *Commerzbank Frankfurt*. Basel: Birkhäuser.

Dryzek, J. S. (1997). *The politics of the earth: Environmental discourses*. Oxford: Oxford University Press.

Feenberg, A. (2002). *Transforming technology: A critical theory revisited*. New York: Oxford University Press.

Feenberg, A. and A. Hannay, eds. (1995). *Technology and the politics of knowledge*. Bloomington: Indiana University Press.

Frei, O. (1991). "Ein ökologisches hochhaus für frankfurt—wenn der bauherr (nicht) will: eine öffentliche stellungnahme." *Die ZEIT*, 15 November, p. 74.

Galbraith, J. K. and M. Berner, eds. (2001). *Inequality and industrial change: A global view*. New York: Cambridge University Press.

Göpfert, C. J. (1991). "Häuser am kaiserplatz gerettet." In *Frankfurter Rundschau*, 25 February, pp. 7–8.
Harvey, D. (2000). *Spaces of hope*. Berkeley, CA: University of California Press.
Herf, J. (1984). *Reactionary modernism: Technology, culture, and politics in Weimar and the Third Reich*. Cambridge: Cambridge University Press.
Hess, D. (2002). *Technology-oriented social movements*. Unpublished manuscript, collection of the author.
Horkheimer, M. and T. Adorno, (1947). *Dialectic of enlightenment*. Amsterdam, Netherlands: Querido.
Lattka, H. (1991). "Drei neue hochhäuser und ein freies plätzchen." In *Frankfurter Neue Presse*, 19 June, p. 15.
Mol, A. P. J. (2003). "The environmental transformation of the modern order." Pp. 203–24 in P. B. Thomas J. Misa, and Andrew Feenberg, eds., *Modernity and technology*. Cambridge, MA: MIT Press.
Naess, A. (1995). "The shallow and the deep, long-range ecology movements: A summary." Pp. 141–50 in G. Sessions, ed., *Deep ecology for the 21st century: Readings on the philosophy and practice of the new environmentalism*. Boston: Shambala.
Pepchinski, M. (1998). "Commerzbank." In *Architectural Record* 186 (1): 69–80.
Rorty, R. (1982). *Consequences of pragmatism*. Minneapolis: University of Minnesota Press.
Rorty, R. (1998). *Achieving our country*. Cambridge, MA: Harvard University Press.
Scherf, D. (1998). "Die kathedralen des kapitals kratzen am himmel" ["The cathedrals of the capital scratch the sky"]. In *Offenburger Tagblatt*, 18 April.
Schreiber, M. (1997). "Wildwest am main." In *Der Spiegel* 38th week (1997): 226–28, 230.
WCED. (1987). *Our common future*. New York: United Nations World Council on Economic Development.
Wentz, M. (1987). "Für neue hochhäuser: rücksicht auf die stadtteile." In *Frankfurter Rundschau*, 10 February, p. 15.

NOTES

1. Portions of this chapter have been published as Steven A. Moore and Ralf Brand, "The Banks of Frankfurt and the Sustainable City," *Journal of Architecture* 8, no. 1 (Spring 2003): 3–24. In general, all translations from the German are by Ralf Brand.

2. The historical data regarding the history of Jewish banking in Frankfurt cited here is derived from installations at the *Judengasse Museum*, Frankfurt am Main, 2 August 2001 and the *Juden Museum*, Frankfurt am Main, 4 August 2001.

3. The legacy of the Marxist interpretation of National Socialism has left us with a distorted understanding of Nazis as rational and capitalist. In fact, Nazis embraced an irrational form of anticapitalism.

4. This argument is supported by contemporary statements such as that attributed to Constanze Kleis, author of *Frankfurt—Das Insiderlexikon* (Munich: C. H. Beck, 1997). "There is still an honest atmosphere [in Frankfurt] that makes contra-

dictions more visible than elsewhere." <www.chancen.net/stadt_und_land/deutschland/frankfurt/urban_village.jsp> (accessed 8 February 2002).

5. This claim is empirically supported by census data reported and interpreted in the local press indicating that "the ratio of foreigners in Frankfurt is 30 percent and thus the highest among all larger German cities. But yet there were no xenophobic attacks." <www.chancen.net/stadt_und_land/deutschland/frankfurt/urban_village.jsp> (accessed 8 February 2002).

6. It is politically significant that the Römerberg was reconstructed as a historic ideal under a CDU mayor (Walter Wallmann) between 1981 and 1986. This project was interpreted locally as a deliberate protest against the loss of local identity. <www.chancen.net/stadt_und_land/deutschland/frankfurt/urban_village.jsp> (accessed 8 February 2002).

7. Serhat Karakayali accused the *Aktionsgemeinschaft Westend* of anti-Semitism because it bluntly accused Jewish real estate speculators of being, at least in part, responsible for the housing problem. The radical *Sponti-Szene* group, that included Fischer, Cohn-Bendit, et al., made clear that their fight was not targeted at Jews but at police, the communal office for foreigners, the judicial system, and so forth. Later Karakayali argued that "the target of the Häuserkampf must be more the SPD and the banks." Serhat Karakayali: *Across Bockenheimer Landstraße. = diskus 2/00.* <www.copyriot.com/diskus/2_00/a.htm> (accessed 8 February 2001).

8. The concept of housing and shops "had been brought about due to 'special encouragement by the city,' said Kohlhausen" (Göpfert, Claus-Jürgen: *185 Meter Büros am Kaiserplatz*). Frankfurter Rundschau, 28 June 1991, page unknown.

9. Apportioning credit for the innovative character of the Commerzbank building design is a quite difficult task. Although the office personnel of Sir Norman Foster certainly deserve much of it, major contributions were clearly made by the engineering firm of Ove Arup, the cladding manufacturer, and many other design team participants.

10. Although there is no official count of housing units displaced by the Commerzbank Tower project, activist Nicole Weidemann has documented a total loss of nearly thirty thousand residents in the banking and West End districts since the late 1960s.

11. "IG Bau" is *Industriegewerkschaft Bau*, or Industry Labor Union for Construction (workers).

FIVE

Story versus Space

> It is space not time that hides consequences from us.
>
> —John Berger, "Ways of Seeing"

Until this point in our investigation, it has been enough to consider how the citizens of Austin, Curitiba, and Frankfurt talk about their aspirations to live sustainably. These local conversations concerning politics, the environment, and technology have helped us to understand the categories of perception that are shared between citizens as well as the vocabularies that separate them. But, as John Berger warns, the history of public talk must be disciplined by geography. There are too many stories told only in spaces that are hidden from us by barriers that remain invisible to ordinary citizens. Or put another way, stories like the ones I have reconstructed in the preceding chapters are so powerful that analysts may be prematurely satisfied with having gotten it "right" (Soja 1989).

In this chapter I argue that public talk shapes not only history but also geography—the physical city itself. A corollary of this argument is to hold that different kinds of public talk should shape cities in different ways. The French geographer Henri Lefebvre argued in 1974 that each society—or better, each "mode of production"—produces its own peculiar kind of space. This is to say that a theocratic society like Iran, a Marxist society like Cuba, or a liberal capitalist society like the United States would each produce spaces in a way that order human relationships in a manner consistent with its political disposition—its assumptions about the proper relationship between individuals and the society as a whole (Lefebvre 1991).

In the context of this study, such logic suggests that Austin space should reflect a liberal-anarchist disposition, Curitiba a liberal-realist disposition, and Frankfurt a liberal-minimalist disposition. But, on the basis of these case studies, it is necessary to add to Lefebvre's insight that political talk does not produce space in a vacuum. Talk about the natural environment into which societies have inserted themselves and the technologies they have used to alter those natural environments is equally important in assessing the local history of space (Moore 2001). It is the porous relationship between politics, technology, and the environment that is so powerful. Toward this end, this chapter first considers how public talk influences city form in general before examining the spaces of specific cities.

Public Talk and City Form

History is full of accounts in which public talk has reordered social space that we now perceive as only natural. Two examples will help to make this point.

The first is what historians refer to as the public health movement, a public conversation that began in mid-nineteenth-century Britain. On one side of this conversation were the religious faithful who argued that the ill health of Britain's poor was God's retribution on those who lived slothful lives.[1] On the other were the Utilitarians, who argued that the destitution of the poor followed disease, not the other way around. Their reasoning was that ill health derived from degraded environmental conditions not moral depravity (Chadwick 1965; Taylor 1996).[2] And from our vantage point in history we know that utilitarian logic (aided by the later addition of the germ theory of disease) has generally been embraced by the faithful as well.

The same vantage point in history, however, tends to obscure how this public conversation altered urban space. Michel Foucault has linked the modern institutions of health (hospitals, clinics, and nursing homes) to those of justice (courts and prisons) through the modern spatial concept of the quarantine (Foucault 1977). The Utilitarian notion that people who threaten the productivity of society should be spatially segregated from it was then a new idea, but one that we now consider only natural. Even if we remain skeptical of Foucault's critique of modernity, though, we would have to recognize that the hospitals, medical districts, and criminal justice compounds that are found in nearly every Western city would not exist without the eighteenth- and nineteenth-century public talk that enabled them. Building types and urban spatial patterns are, then, the reification or materialization of the public talk and institutional agreements that precede them.

A second example of public talk that has reordered space is the environmental movement. Before John Muir (1838–1914) began his campaign to preserve particular threatened natural environments in the American West, we North Americans had not yet developed a distinct category of space we now see as normal—the nature park. Beginning in the late nineteenth century, at roughly the same time as talk about public health gained steam under the pressures of the Industrial Revolution, successive government administrations followed the lead of President Theodore Roosevelt in drawing spatial boundaries between what we understand to be undisturbed "nature" and what was destined to become nonnature. Since the time of Roosevelt and Muir, the North American landscape has virtually been transformed in accordance with the linguistic categories of public talk in that era.

I have previously argued that these two kinds of public talk—about the public's health and the environment's health—were initially isolated because the vocabularies of the two groups were simply allergic to one another—it was impossible for them to hold a civil conversation because their political dispositions were so different. They became fused, however, as a single conversation about sustainable development in the late twentieth century, a hundred years after they were initiated, because of changed social and environmental conditions (Moore and Engstrom 2005; Beck 1992). This is to argue that our spatial categories change in response to, and as a reflection of, our social and linguistic categories. Or as Derek Gregory and John Urry put it, "Spatial structure is now seen not merely as an arena in which social life unfolds, but rather as a medium through which social relations are produced and reproduced" (Gregory 1985). This is why we can now have twenty-first-century institutions in which hospitals have large gardens and nature centers have therapeutic suites. As the boundaries of public talk become blurred and change, so do public spaces.

There are, of course, many other examples of public talk that have reordered space—industrial talk that led to urban zoning, sports talk that has privatized the spaces of public competition, or race talk that has segregated human populations. In the context of the cases studied here, Frankfurt's *Judengasse* is certainly the most extreme example. All of these examples are helpful, but perhaps redundant. We might, then, simply say that "(social) space is a (social) product" (Lefebvre 1991). This logic recalls that in chapter 1 and suggests that space evolves—that there are historical stages of spatial development of which the sustainable city is but one in a long history of other conversations.

Sustainable Urban Form

David Harvey acknowledges that "the concept of 'sustainability' . . . though easily co-opted, points to spatiotemporal horizons different from those of capital accumulation" (Harvey 2000). In this sense, sustainability is the creation of the "heterotopic" space that Foucault had theorized if never experienced. By this term he meant, I think, space that would reveal rather than conceal the social processes that lie behind it. That space might reveal the social, as well as the material, processes of its making provides a creative opportunity for (and places a great responsibility on) expert designers.

The speculations of Harvey, Lefebvre, and Foucault are highly abstract, but among contemporary planners there has been far more concrete speculation about how the sustainable city might be organized. Peter Newton, for example, has digitally modeled five urban form types to determine which would be more sustainable. He limited the variables studied, however, to alternative land uses and transportation configurations. He concludes that "urban form does matter. . . . [A] trend toward the more compact city, however defined—compact inner city, edge city, corridor city, and so on—will lead to significant environmental improvement" compared to the "dispersed city" in terms of air quality and related phenomenon (Newton 2000). This finding is, however, still abstract and of little use unless we plan to construct new cities in previously uninhabited regions—a prospect that seems highly unlikely in our current situation. A more valuable type of research would be to study how existing urban conditions, the product of historically woven story lines, might be improved.

Katie Williams and her colleagues agree that urban form is significant in determining sustainability, yet they argue, "instead of searching for one definitive sustainable form, the emphasis should be on how to determine which forms are suitable for any given locality" (Williams 2000, 7). In other words, ideal urban forms do not exist in a vacuum disengaged from particular social and ecological conditions. Limited to the considerations of travel distances and fuel economy it takes only a fairly simple quantitative calculation to demonstrate that the compact city is the most efficient; however, when planners consider the variables associated with ecology, social conditions, and economy, research may well lead in other directions.

Others argue that there are "natural spatial laws" at work that limit the ability of citizens to optimize spatial relationships between whatever variables one might want to include. This *structuralist* approach to sustainability holds that settlement patterns result from a dialectic relation between local and global "rules." At times local rules lead to "well defined global forms," whereas at other times settlement patterns "might be understood as the global product of different local rules" (Hillier 1984). In either case, the emphasis is on finding "spatial laws" rather than finding new and better ways to live. A good concrete example, which challenges the structuralist approach, is the desert city. Through structuralist lenses the tightly woven fabric of, say, Isfahan, Iran, with its winding alleyways and system of courtyard houses joined by party walls is a highly efficient response to the city's ecological situation. By creating shade and barriers to desert winds this urban pattern sustains life in a particularly elegant way. Why, then, is Los Angeles—with nearly identical ecological conditions—so different?

Although the interplay between local and global patterns is indeed an important consideration, the construction of a priori rules may tell us more about the rule makers than the conditions of particular places. For this reason I am inclined to place more faith in the history of space and the public talk that we have investigated thus far. This is not to say that global structures do not exist or are insignificant, but that the study of how global patterns are implemented locally is more important than the abstract models themselves. We will return to the relationship between conceptualization and implementation in chapter 6, but for the moment it is enough to argue that utopian ideals of sustainable urban form, disengaged from either local ecological or social conditions, obscure more than they reveal.

But having dutifully warned the reader against acts of abstraction, the next five sections of this chapter will study our three cities in exactly this way—through the quantitative methods of Geographic Information Systems (GIS). Maps are powerful because they edit reality—they erase the information that is in the judgment of the map maker unimportant or distracting to the project of getting from place A to place B, or unimportant in the project of describing property. This observation, of course, raises the question of who gets to decide what is and what is not important. As we shall see, such judgments can be highly controversial and have lasting consequences for how a city is lived. Lefebvre argued that any abstraction of space, especially maps, has political motives. Once the particular ecological or human qualities of space are reduced to mere quantities, it becomes far too easy to substitute "exchange value" for "use value" or to think about rent rather than dwelling (Lefebvre 1991). The analysis that follows is no different. Nor should the reader imagine that because the maps constructed here are derived from census data, or other seemingly objective sources, that they are somehow more correct than the qualitative stories reconstructed from interviews in the case studies. Hardly. These two kinds of stories—one prose and one graphic—are, however, complementary precisely because they see reality differently.

One rationale for such abstract analysis is to reconstruct selected characteristics of cities that are identified with the concept of sustainability to determine whether there are consistent patterns that make them similar. This is an inductive approach—the flip side of trying to determine ideal forms based on deductive principles. Toward that end, the following sections consider, in turn, the evolution of development, the spatial distribution of parks and open space, access to public transportation, population density, and income distribution for our three cities. These categories of analysis are not comprehensive—many other categories such as crime, energy usage, or the distribution of environmental risks might have been used. I selected these only because they are generally considered to be good indicators of urban sustainability and the data are generally, but not always, available, as we shall see in the case of Frankfurt. We shall also see that these categories of analysis are not equally important, yet each one has significant consequences for citizens.

5.1 EVOLUTION OF DEVELOPMENT

In modern Western democracies land-use zoning is the primary tool available for municipalities to manage their spatial development. This broad statement, however, fails to capture the differing legal foundations for municipal authority to regulate space that is a function of their respective federal and state constitutions. This is to say that the implementation of zoning in almost every city is different. In the United States, for example, the concept of zoning was introduced in Modesto, California, in 1880 as a means to remove Chinese laundries from white neighborhoods. By 1913 zoning was embraced in New York City and, when challenged, was upheld by the U.S. Supreme Court in 1916 under the police power of the state to safeguard the "public health, safety, morals, and convenience." On the basis of this history many argue, though not without controversy, that the purpose of zoning in the United States can be associated with the preservation of private property values and not the public value found in long-term land-use planning. This legal foundation is significant in our analysis because it distinguishes the method of zoning implementation employed in Austin and Curitiba from that employed in Frankfurt, where zoning is legally authorized on the basis of long-term ecological planning and landowner compensation. Implementing zoning through the use of police power (which is weak in the United States but strong in Brazil) and through regional land-use planning authority (which is the general case in Europe) logically leads to different spatial patterns (Hall 1996). The significance of this observation is that the means of implementing an idea may influence outcomes more that the idea itself.

Figures 5.1, 5.2, and 5.3 illustrate the evolution of spatial development in Austin, Curitiba, and Frankfurt, respectively. There are two possible interpretations of the data provided by the cities. The first assumes that the data are correct, and the second is more critical.

In the first interpretation, the spatial progression of development over time in Austin and Frankfurt seems similar to, but distinctly different from, that of Curitiba. That Curitiba's development pattern is so different can be interpreted to be the result of the strong police power held by the planners at IPPUC, the quasi-autonomous regional planning agency. The very clear concentric pattern of growth suggests that someone was in control of the development of peripheral land, especially when compared to

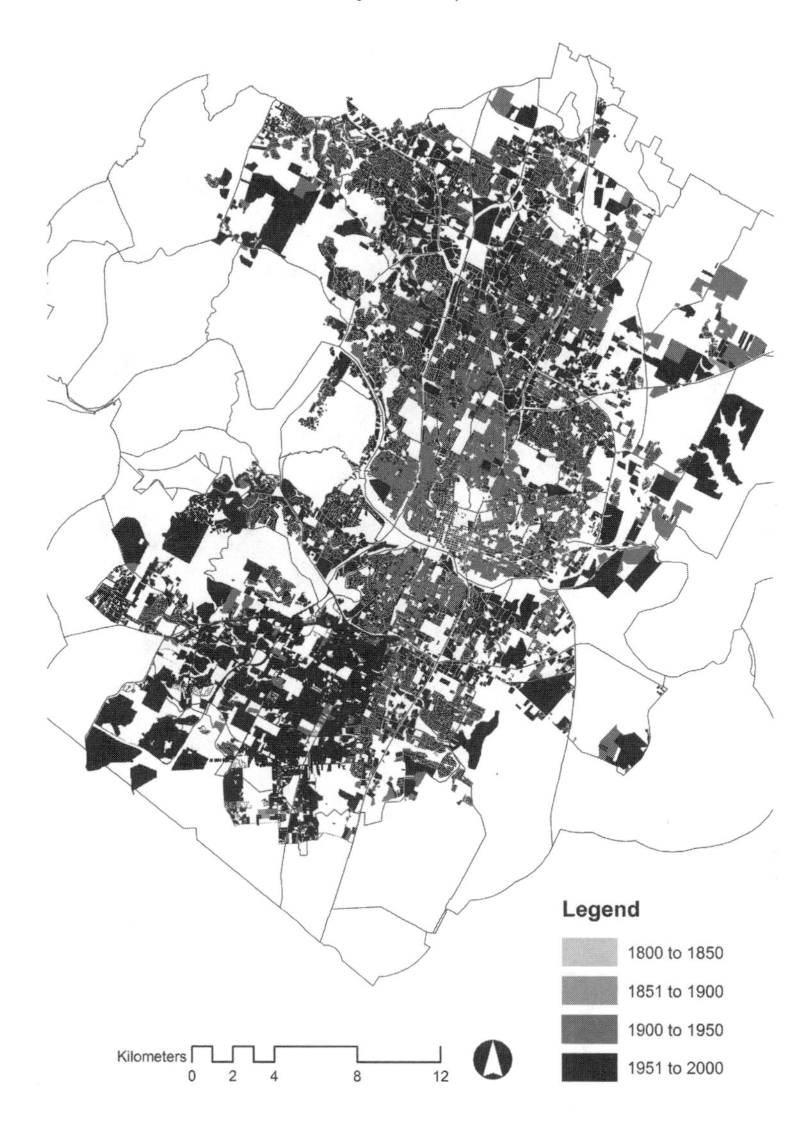

Figure 5.1. Evolution of Development, Austin

the far more ragged, or "leapfrog," patterns illustrated in the maps of Austin and Frankfurt. This comparison might suggest that development patterns of liberal-anarchist regimes (Austin) and liberal-minimalist regimes (Frankfurt) are the same. Closer analysis, however, suggests that the leapfrog pattern of Austin is a function of four factors: (1) the absence of land-use zoning outside the city's current boundary; (2) the desire of

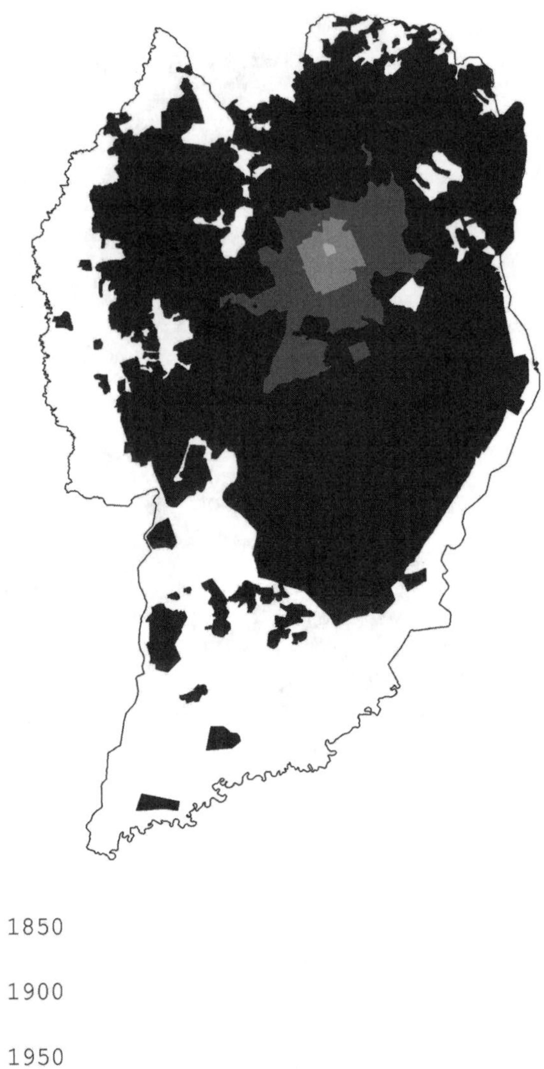

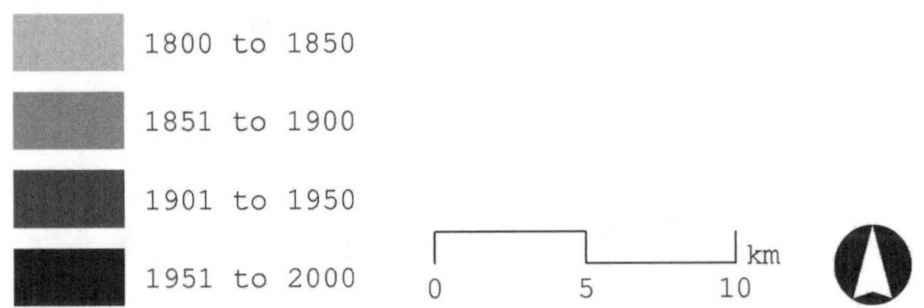

Figure 5.2. Evolution of Development, Curitiba

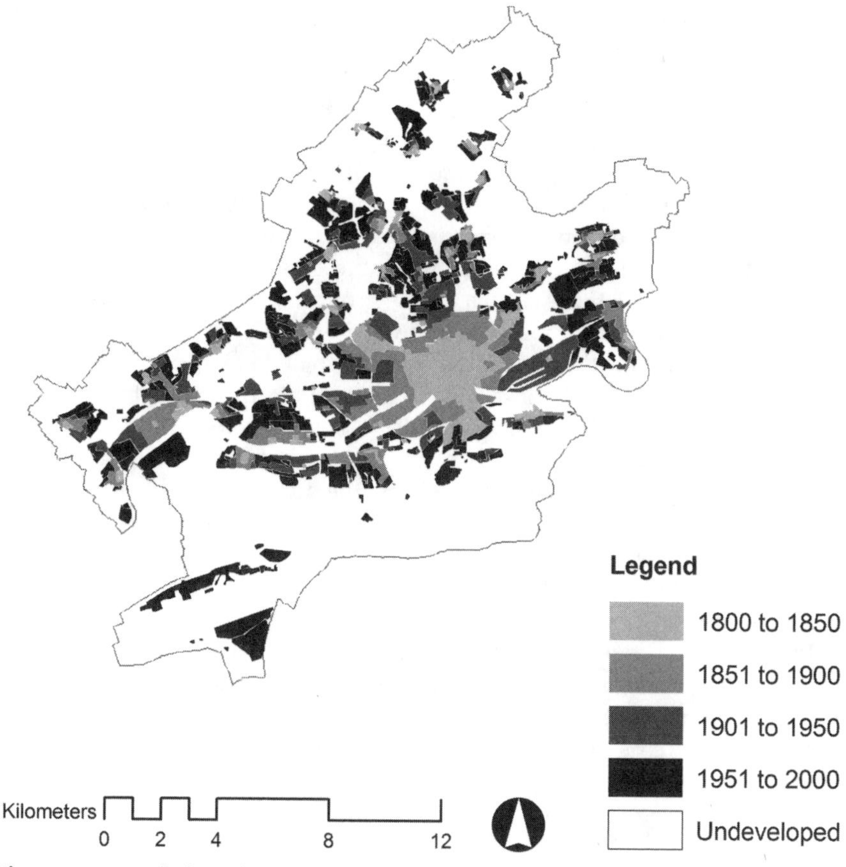

Figure 5.3. Evolution of Development, Frankfurt

real estate interests to buy the cheapest land available within the gravitational pull of the city; (3) the legal ability of developers to create their own municipal utility districts outside city or county jurisdiction; and (4) the very weak policing power available to the city. In contrast, the development pattern of Frankfurt seems to be a function of planners setting aside public parks and open space as population growth required the urban growth boundary to expand. This interpretation is supported by the next set of maps in section 5.2, which analyzes the distribution of public parks and open spaces.

The second interpretation of these maps is based on skepticism of the data provided by Austin and Curitiba. In the case of Austin, an on-the-ground survey indicates that many land parcels indicated as "undeveloped" were obviously developed at the time when background data were collected. The discrepancy indicates holes in the data source. In the visual terms of the map, there is no distinction made between parcels having no records and parcels having no development. In other words, the leapfrog pattern of development accurately reflects what is on the ground in some locations but not in other. In the case of Curitiba, the map is suspect for exactly the opposite reason—it is too regular and not backed by reliable data. Rather, it seems that an IPPUC technician simply approximated the edges of development in a particular period on the basis of institutional goals rather than demographic facts. In the case of Frankfurt, however, the data appear to reflect what is on the ground and is backed by sophisticated data. This second interpretation is more satisfying.

My point in replicating maps that I know to be at least partially unreliable is not to question the integrity of municipal governments but to argue that these maps do accurately reflect the stories being told by the respective dominant regimes in each city if not what is on the ground. The somewhat chaotic map of Austin reflects the atomization of its citizenry. The tightly controlled and nearly concentric rings of Curitiba reflect technocratic authority. And the leapfrog pattern of Frankfurt accurately reflects the agricultural land and open space that has been integrated into the cityscape.

In sum, the evidence here suggests that different kinds of public talk do lead to different development patterns. This is to say that viewed through the digital lenses of geographic information systems the evolution of development patterns of Austin and Frankfurt may appear to be similar, especially in contrast to Curitiba, but are lived very differently.

5.2 PUBLIC PARKS AND OPEN SPACE

Each city classifies undeveloped green space available for public use differently. Some include cemeteries, bodies of water, athletic fields, and undeveloped municipal property and others do not. Figures 5.4, 5.5, and 5.6 of Austin, Curitiba, and Frankfurt, respectively, are an attempt to compare apples to apples.

This set of maps suggests that both Austinites and Frankfurters enjoy a substantial amount of parks and open space and that it is fairly well distributed—meaning that it is accessible to citizens in many neighborhoods. Although the data do not support it, the city of Curitiba describes its own situation like this:

> Curitiba has 26 parks of well-preserved environment, with rich and diversified fauna and flora. The city has a green area of 55 m^2 for each resident and is known as the Ecological Capital of Brazil. (IPPUC 2005)

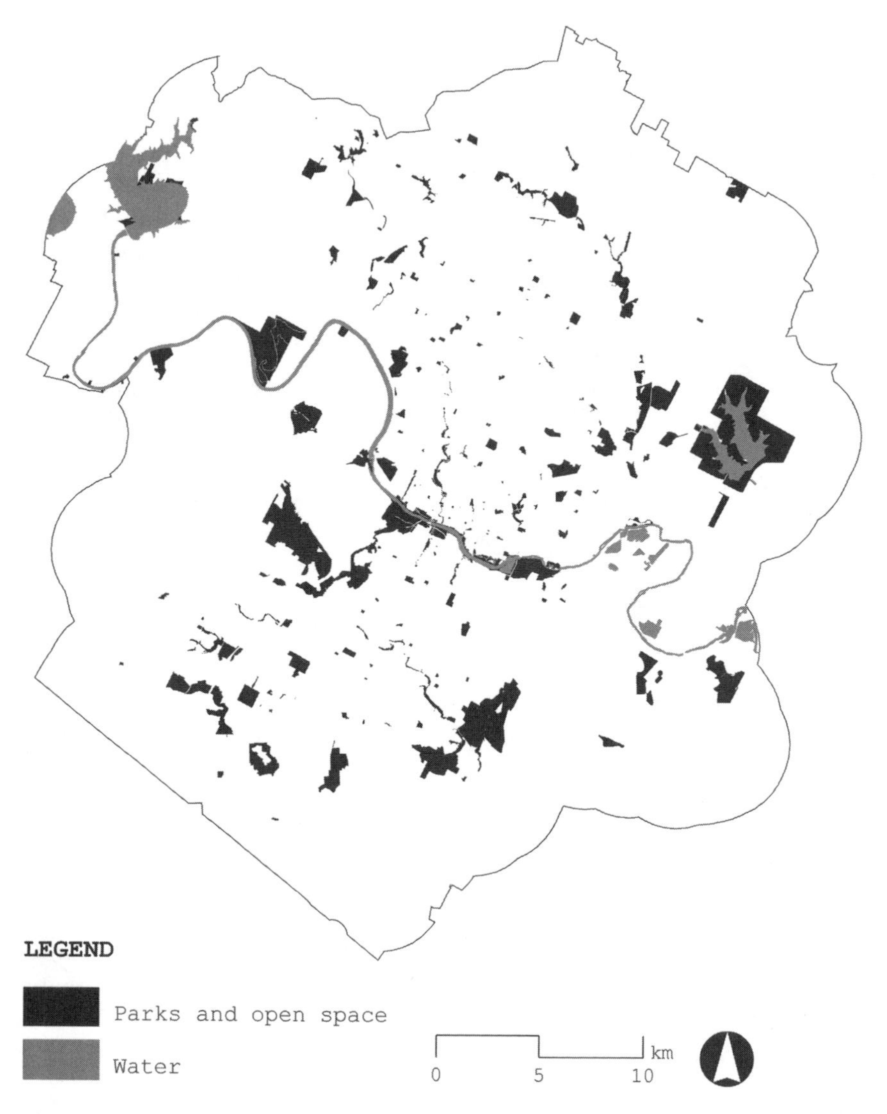

Figure 5.4. Public Parks and Open Space, Austin
Courtesy of the author.

The problem here is that when I calculated parks and open space for each city using the same criteria, the data provided indicated that Austin had 68.39 square meters per person, Curitiba had only 9.32 (not 55) square meters per person, and Frankfurt 97.79 square meters per person. These totals create a ratio of 7.3 : 1 : 10.5 respectively. We could discount Curitiba's

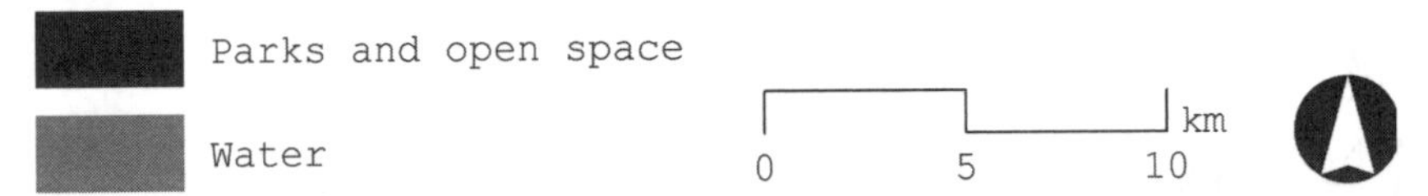

Figure 5.5. Public Parks and Open Space, Curitiba
Courtesy of the author.

claim as an example of the propaganda discussed in chapter 3, or we could understand it as a cultural difference in what is considered accessible open space. No matter the reason for the discrepancy in absolute numbers, the graphic comparison tells the story best: even if the citizens of Curitiba believe that they have good access to parks and open space, the experience of citizens in Austin and Frankfurt is lush by comparison.

The Curitiba map also reveals another problem—the spatial distribution of the small area of parks and open space that does exist. The pattern

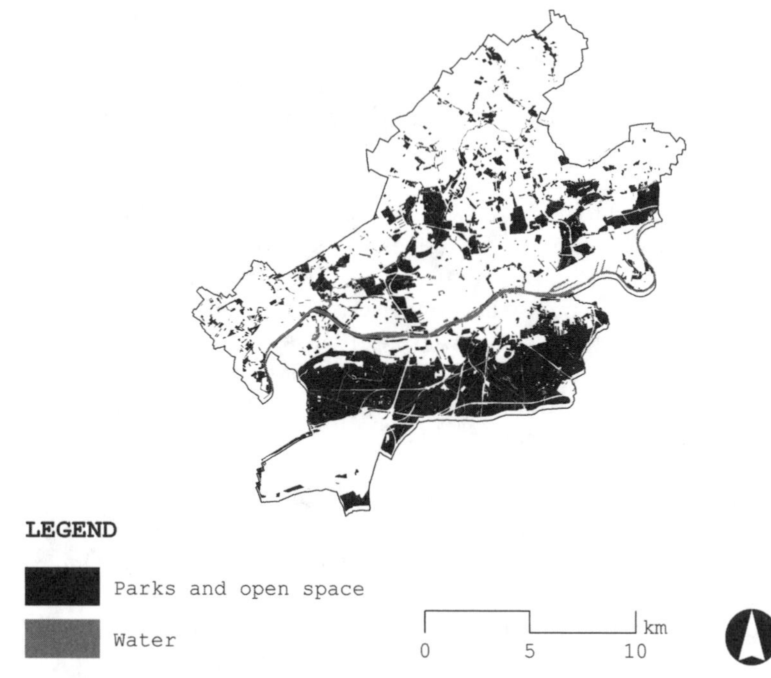

Figure 5.6. Public Parks and Open Space, Frankfurt
Courtesy of Planungsverband Ballunsraum.

of open space documents that the wealthier and less dense northern residential districts are better served than the poorer and denser southern residential districts.[3] This is not exactly unexpected—taxpayers get served first—but more important is that the distribution of open space seems to have no correlation to the extremely high densities along the city's structural axes (illustrated in figure 5.8) where it is most needed. The pattern suggests that decisions were made to optimize developer profits rather than citizen health and welfare.

Developers, of course, are prone to use the logic of environmentalists in arguing for greater urban density. Although both groups may agree on this point in principle, environmentalists also argue three points in favor of the environmental services provided by parks that indirectly reduce density by taking up space. First, parks significantly reduce the urban "heat island effect"—the buildup of heat and airborne particulate pollution over cities that is undesirable because of the associated medical and cooling costs (Baker 2001). Second, the psychological benefit enjoyed by urban dwellers in proximity to parks makes even higher densities tolerable, thus overcoming the seeming "inefficient" use of expensive land. And third, the existence of park flora attracts fauna, like the famous peregrine falcons of New York's Central Park that feed on other undesirable but incorrigible urban species such as rats. Curitiba, in particular, has done a good job in making recreational land double in function as storm water retention areas, thus providing a vital ecological service to the city.

In 2000, Austin's citizens voted overwhelmingly to acquire more than 17,000 acres of land to the southwest of the city and on top of the Edwards Aquifer that was threatened by development. Most of that planned acquisition does not yet appear in figure 5.4. The largest area without public open space is, of course, in the western hills, dominated by the most expensive homes where open space is private space. Another example of the public affinity for open space in Austin is that the very heart of the city is an open green space that surrounds Town Lake and which is home to the city's much beloved hike and bike trail. It is fair to say that this center cityscape is paradoxically nonurban.

In contrast to Austin's parklike core, the edges of the Main River in central Frankfurt are far harder and more urbane. In lieu of a hike and bike trail, it seems that Frankfurters prefer cafes and hardscape in their city center. But Frankfurters, as do Germans in general, also value the idea of a public forest, just not in the city center. Going for a walk in the public forest south of the Main River is not, however, a substitute for public open space of other kinds. Rather, it is a social habit unfamiliar to Texans or Brazilians.

In sum, I argue that the public talk in all three cities has produced park and open space that is highly valued by citizens, but that Curitiba has been less successful than the technocrats of IPPUC claim and less fair in distributing space equitably.

5.3 PUBLIC TRANSPORTATION

The three cities under investigation do not have the same, or even similar, transit systems. Austin has only buses and taxis, but in 2005 voters passed a referendum funding a phase one commuter rail system. Curitiba, of course, has its well-known RIT bus system that includes routes that operate at four different scales along its structural axes and radially between them. And Frankfurt has access not only to a regional rail system but also an efficient subway and local bus service. These different systems are illustrated in figures 5.7, 5.8, and 5.9, respectively.

These maps indicate the geographic routes of various transit types, not ridership quantities, so they may be slightly deceiving. In purely quantitative terms, Austin reported a total annual public transit ridership of slightly more than 34 million, or about 52 citizen trips per year (Metro

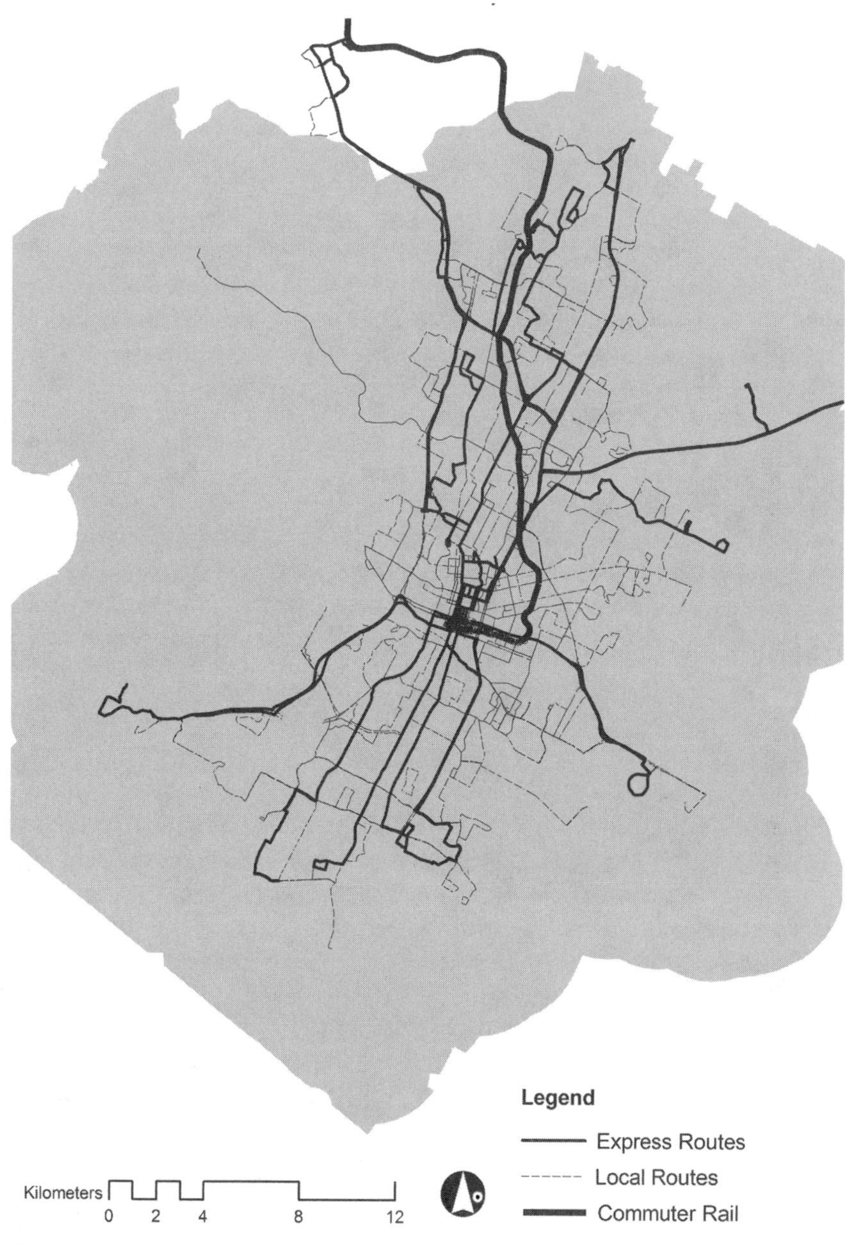

Figure 5.7. Public Transportation, Austin

Figure 5.8. Public Transportation, Curitiba

Figure 5.9. Public Transportation, Frankfurt
Courtesy of traffiQ Frankfurt am Main.

2005); Curitiba 730 million, or about 456 citizen trips per year (IPPUC 2005); and Frankfurt 219 million, or about 336 citizen trips per year (Kassen- und Steueramt 2005).[4] These numbers speak for themselves—Frankfurters ride public transit 6.5 times more frequently than Austinites, and Curitibanos 8.8 times more frequently. We can reasonably conclude that Austin, unlike Curitiba or Frankfurt, is automobile dependent and largely suburban.

Figure 5.10. State of Texas Automobile License Plate
Courtesy of the author.

These three maps indicate, then, not the quantity of service provided but the spatial distribution of that service. Even a quick visual analysis suggests that citizens of Curitiba will find it hard to avoid public transit and that those of Frankfurt will find it easy and efficient to use. Perhaps most striking about Frankfurt's transit route pattern is its central position as a hub of regional passenger rail service—a condition certainly consistent with its regional economic role. In stark contrast, only some citizens of Austin have access to any type of public transit at all. The most interesting aspect of Austin's transit network is that it is available to neither the very rich (who live in the hill country to the west of the city) nor the very poor (who live in the black loam prairie to the east of the city). From a planning perspective, one might argue that neither area is viable for public transit because of their low population density, a point that will be illustrated in the next section. Another interpretation, offered by East Austin residents, is that wealthy whites want to minimize public access into their neighborhoods and poor people of color are denied it. This interpretation will become clearer in the last section of this chapter.

In sum, the analysis suggests that the dominant story line constructed by Frankfurters (progressive capitalists) and that constructed by Curitibanos (technocrats) see public transportation to be an important part of the urban future. In contrast, the rugged individualists of Austin campaigned that "it costs too much and does too little." This attitude is nowhere better illustrated than on the state of Texas automobile license plate (figure 5.10), in which an ideal citizen is pictured riding his horse alone at night across the deserted prairie.

5.4 DENSITY

The advocates of transit-oriented development (TOD) generally argue that tying public transit routes to zoning that permits increased population density is a key "best practice" of sustainable urban development (Pinderhughes 2004). This logic seems intuitive—relatively high densities are required to make public transit economically viable, and correspondingly, both automobile emissions and travel times are most effectively reduced by the availability of public transit. If the TOD hypothesis is a valid one, we would expect to see, in the most sustainable cities, a pattern that correlates public transit routes to increased density. Figures 5.11, 5.12, and 5.13 test this hypothesis. In broad terms, the average population densities for our three cities are: 16.9 persons per hectare for Austin; 102.5 persons per hectare for Curitiba; and 63.2 persons per hectare for Frankfurt. Using Austin as the base, a ratio of overall density for the three cities is 1.0 : 6.0 : 3.7—a rather surprising spread for cities making similar claims.

Contrary to the TOD hypothesis, the pattern in Austin is one of overall low density, with very little correlation of elevated density to public transit routes. The map of Curitiba, however, indicates exactly the opposite—that there are areas of very high density that generally follow the structural axes of public transit discussed in chapter 3. Thus, Curitiba is clearly an example where what we now refer to as TOD has been employed as a conceptual planning tool for nearly a half century. Finally, Frankfurt, too, illustrates a pattern in which transit and density are correlated, if somewhat less so than in Curitiba. The advocates of TOD would, then, quickly select Curitiba as the most and Austin as the least sustainable city of the three. This judgment would, however, ignore more elusive social criteria.

In response to this challenge, there is evidence collected by others to suggest that greater density generally supports social equity. Elizabeth Burton, for example, argues that not only are services more likely to be available to the poor in compact cities as opposed to rural locations but that social and economic opportunities increase in proportion to higher density (Burton 2000). In the stories reconstructed in chapters 3 and 4, concerning Curitiba and Frankfurt, I documented that public talk about the relationship between transit and economic opportunity was explicit and supports Burton's general argument.

This logic, however, is contradicted in Curitiba in at least one respect. Some of the least equitable spaces in Curitiba, the *favelas*, or informal settlements located in figure 5.14, are quite dense, especially by Austin's sparse standard. The *favelas* do not reach the high densities found along the structural axes of Curitiba because these citizens, who build their own modest dwellings, lack the technology to build so tall. This is another case where walking through these neighborhoods would produce a different

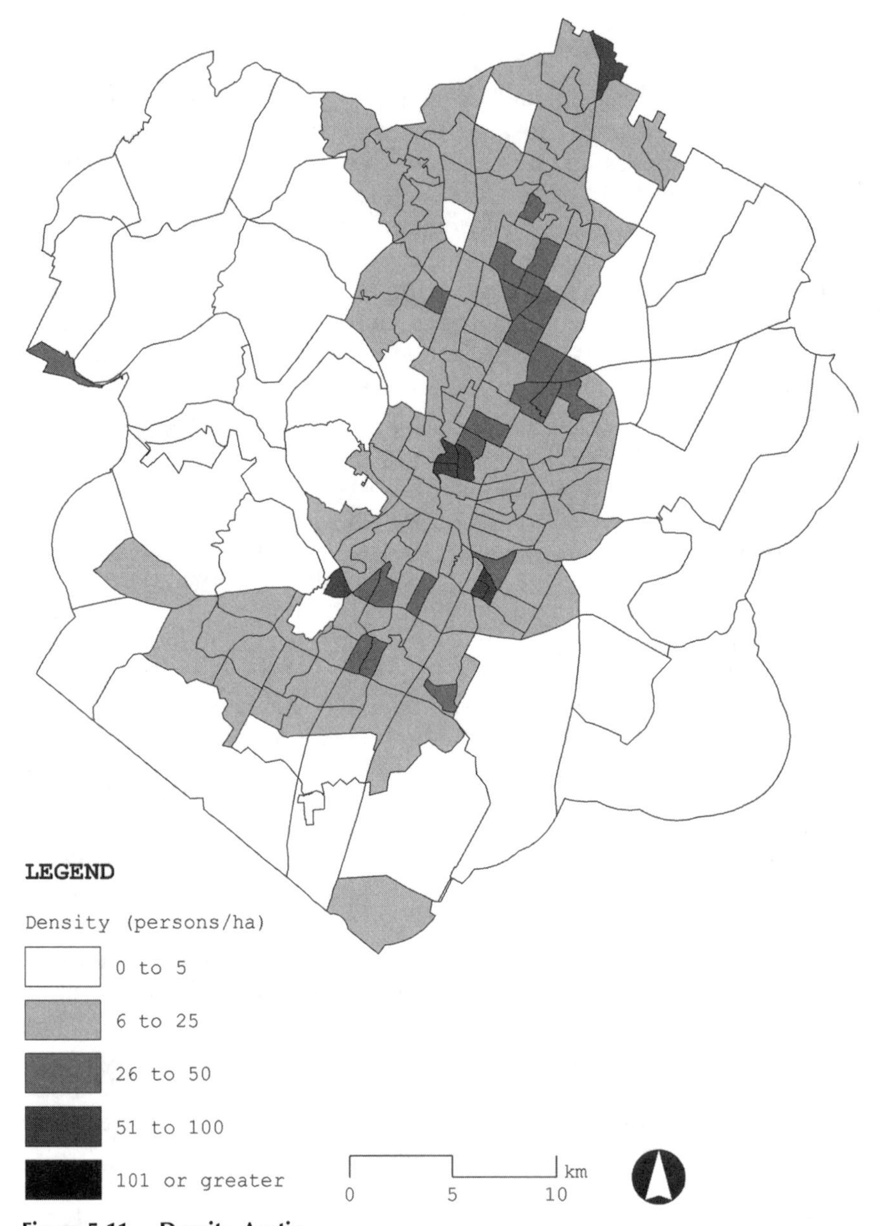

Figure 5.11. Density, Austin

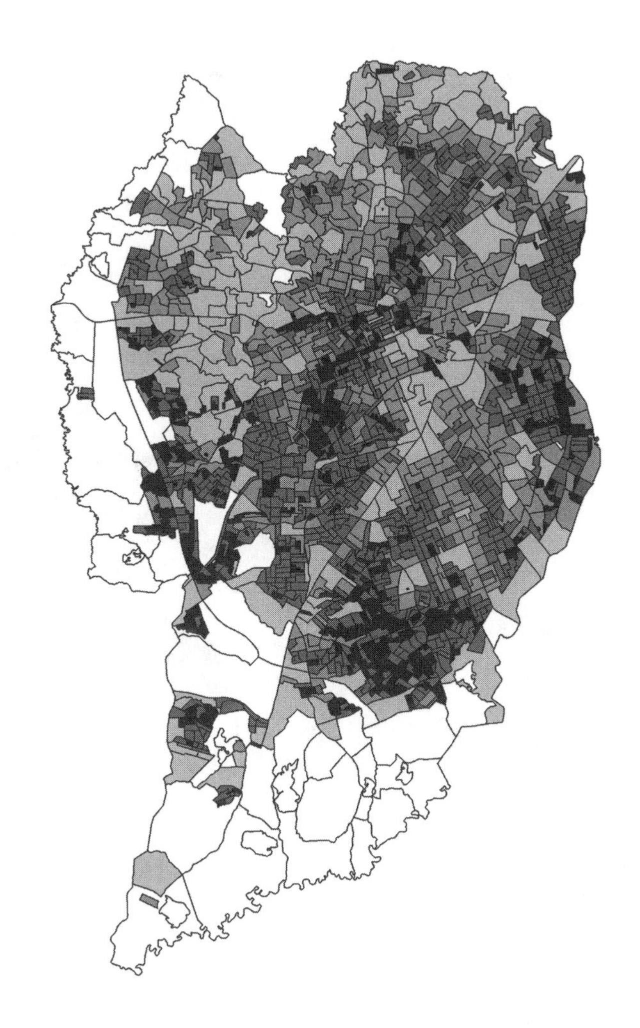

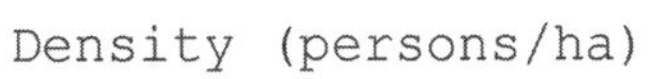

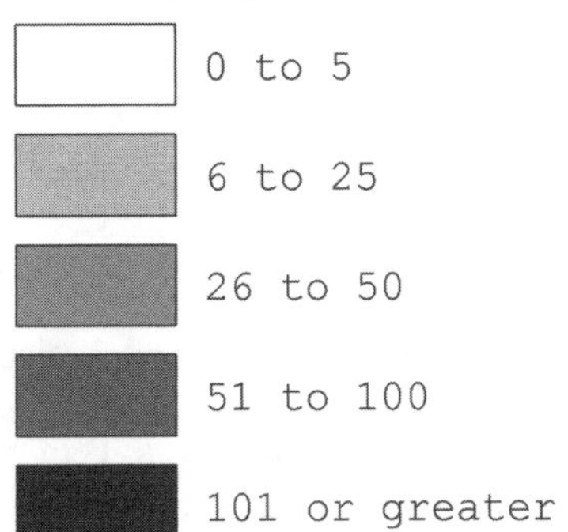

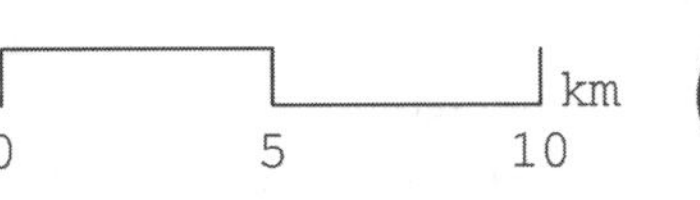

Figure 5.12. Density, Curitiba

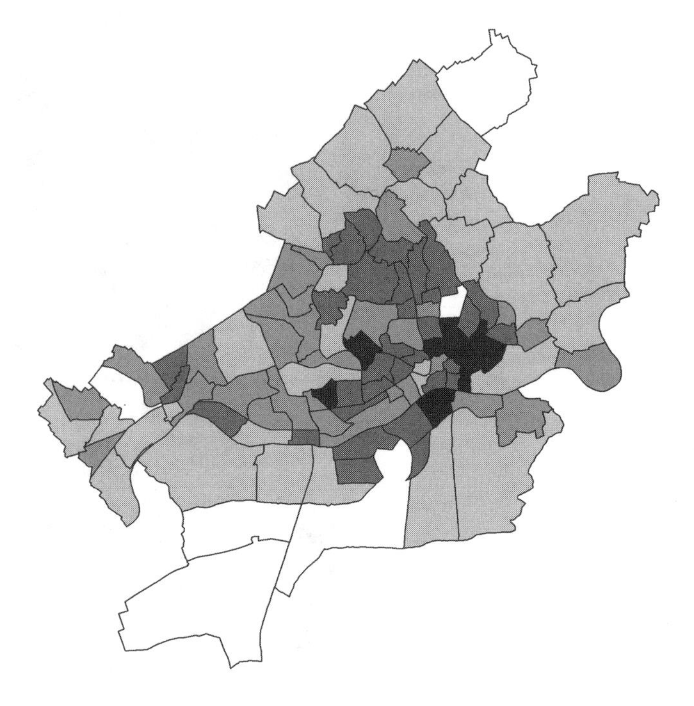

LEGEND

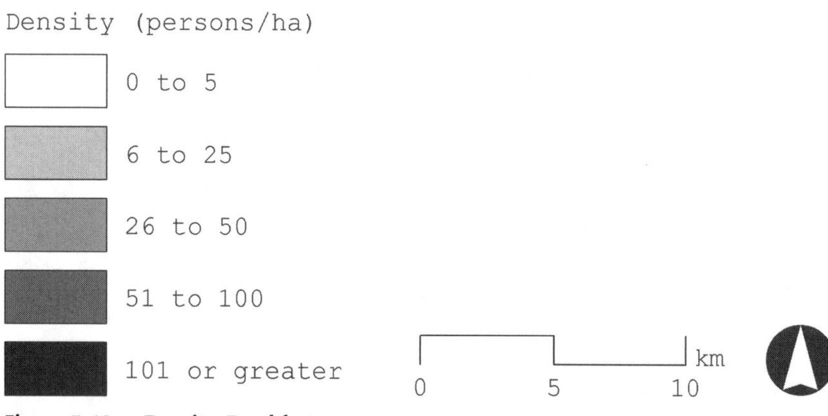

Figure 5.13. Density, Frankfurt

assessment than looking at the map of the same space that illustrates only relative density. Equivalent densities near the structural axes are very middle class in comparison to the obvious poverty of the *favelas*. In other words, equal densities may not live the same. Nonetheless, it is fair to say that the availability of public transit to *favela* residents does provide access to jobs, and thus some equity, even if higher densities in general cannot always be correlated to increased social equity.

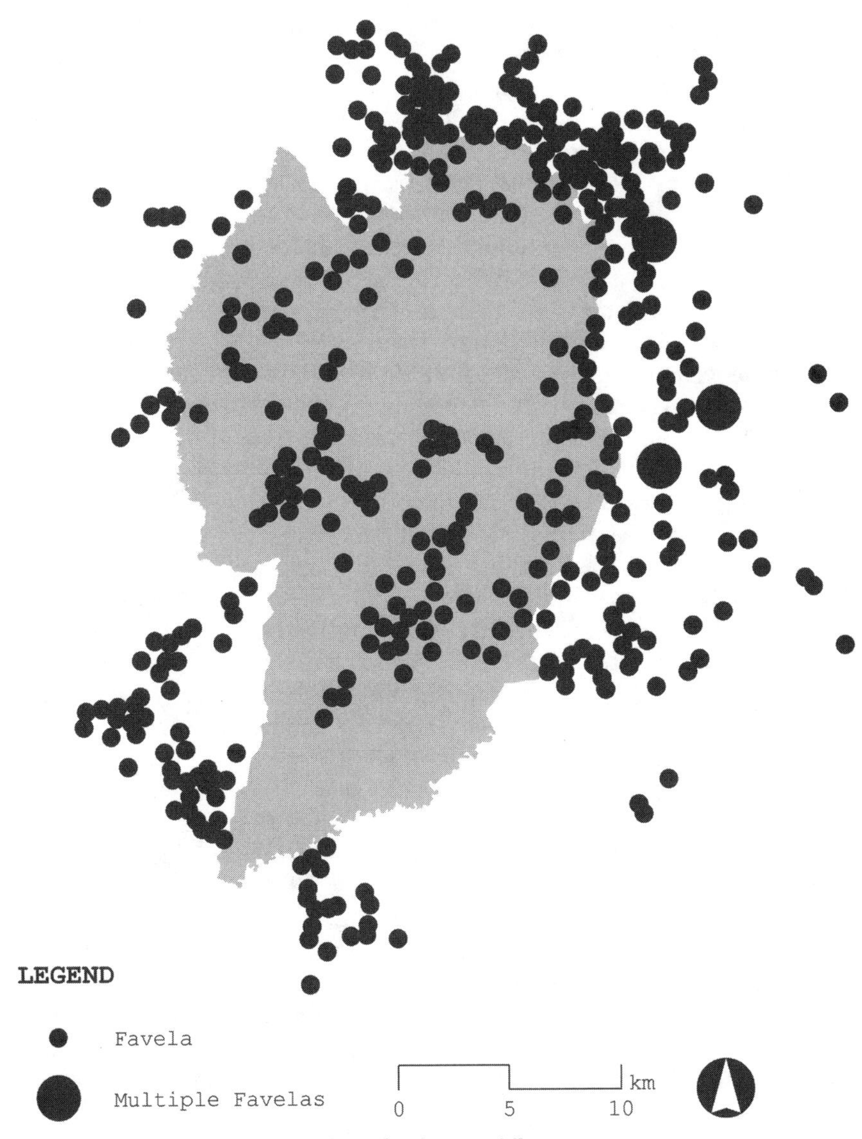

Figure 5.14. Informal Settlements (*favelas*), Curitiba

Burton's empirical findings are also consistent with the more philosophical argument made by Richard Sennett (Sennett 1998). In Sennett's view democratic tolerance, public talk, and equity can flourish only in those spaces where citizens are constantly in contact with other citizens who are not like themselves—who speak differently and tell different stories. In the limited sampling of our three cities this logic too holds up to scrutiny. In the very low density found in Austin, Sennett would expect to find the least amount of contact and conversation between citizens of differing backgrounds if only because with more space available people can avoid each other. This is especially true where there is no effective public transit that serves all classes of citizens. The evidence in Austin in particular, as well as Curitiba and Frankfurt, supports his a priori claim. This conclusion will become clearer in this chapter's final section, on income distribution.

The evidence collected from our three cities, however, does not uniformly find benefit in increased density. The generally negative impacts that citizens associate with increased density are also documented in the literature (Williams 2000). In the three cases studied here, wealthy Austin citizens mentally associate higher densities with crime and loss of privacy while poor Austinites associate higher density with gentrification and the dislocation of their communities. This negative association also holds true for Curitiba's most informed citizens, such as Rodolpho Ramina, who consistently complain of noise and air pollution along the structural axes. Only in Frankfurt have some citizens who initially rejected increased density actually come to embrace it as being in their interest. In chapter 4, I documented the turmoil that erupted in Frankfurt when the *Westend* neighborhood was threatened by increasing density and mixed use in the 1970s. The *Häuserkampf*, or housing struggle, that ensued was eventually put to rest by twenty years of public talk that transformed public opinion by developing new local codes that satisfied the seemingly contradictory needs for increased density, mixed use, and environmental preservation. The point here is that the long-term public talk of Frankfurters transformed an existing neighborhood into a new kind of urban space not found in either Austin or Curitiba. Through talk, code making, and experimentation with many different kinds of possible solutions over the years, the citizens of Frankfurt (with the help of its expert architects and planners) finally convinced themselves that a very particular kind of higher density was the best solution.

In sum, it is reasonable to argue that low densities are consistent with the story lines constructed by the rugged individualists of Austin but not with the general interests of sustainable urban development as they are more successfully constructed in Curitiba and the European Union. It is also fair, however, to say that most people will reasonably resist higher density until they can convince themselves that it is in their own interest.

5.5 INCOME DISTRIBUTION

In each of the preceding sections of this chapter—the evolution of development, public parks and open space, public transportation, and density—I have reconstructed spatial realities in the three cities under investigation that compete with the stories told by citizens in chapters 2, 3, and 4.

Nowhere, however, were the stories told by locals and the geographic data in greater competition than regarding income distribution. Even the difficulty I experienced in gathering income data supports the notion that information about its distribution is often suppressed in public conversation. People, unless asked specifically, simply do not like to talk about the spatial consequences of income inequity—it is uncomfortable. The lack of public talk about income distribution can, then, be disciplined only by the presence of data from other sources.

In Austin, income distribution data were readily available from census information. In Curitiba, however, it took more effort. During Lerner's second term as governor, income distribution data for the city was reportedly made unavailable but by 2005 was made only difficult to get. In Germany, such data are simply not available from public sources by federal statute, apparently on the grounds of privacy. German data are available from private sources, but I have not included it here because of its prohibitive cost and the legal limitations placed on its use.. What follows is a reappraisal of the dominant and counter story lines told by Austintes and Cutibanos. A more lengthy analysis is required in this section because the available spatial data suggest that, for Austin in particular, there was a third suppressed story line that will alter our understanding of the stories told by rugged individualists and environmentalists. Telling that story requires new voices.

Austin's Income Distribution

Figure 5.15 shows the spatial distribution of median household income in Austin in U.S. dollars for 2000. The income groups mapped can be understood as classes of people, and the pattern clearly illustrates that social classes are spatially separated by Interstate Highway 35, which runs north and south through Austin, roughly parallel to the Balcones Fault—the geological fault line that separates the rugged hill country to the west of the city from the black loam prairie to the east. The fault line can be understood, then, in economic as wells as geological terms. Were I to superimpose race distribution data on this map of income distribution, the resulting map would also show that there is a strong correlation between living in east Austin with low income and being a member of a minority community that is predominantly African American or Latino. This is to say that Austin is an economically and racially segregated city.

To the citizens of Austin this is hardly news, but it is not a condition that gets talked about much. Austin is, after all, a southern city that was founded in the antebellum period. It does not take much investigation to confirm that such conditions are common in the region and have existed for a long time. What is more surprising is that the conditions of spatial segregation have lasted so long. Other American cities, even northern ones—Dallas, Chicago, and Detroit, in particular—also suffer from segregation,

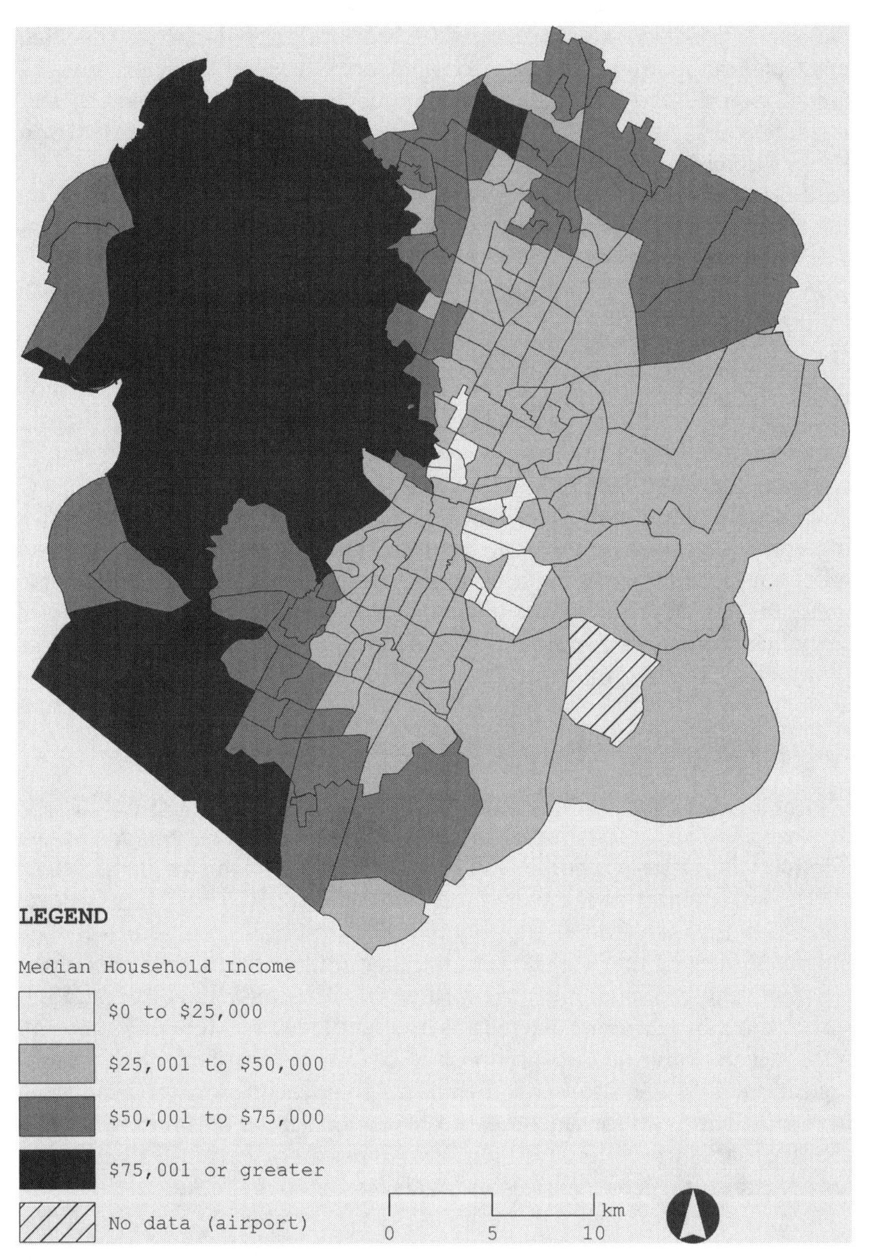

Figure 5.15. Income Distribution, Austin

yet none of these lay claim to the progressive politics that are so at home in Austin. In this context, Austin is something of a perplexing anomaly.

In chapter 2 I mentioned in passing that Austin's first attempt at planning occurred in 1928. A portion of that document, which I quote at length, reads as follows:

> There has been considerable *talk* in Austin, as well as other cities, in regard to the race segregation problem. This problem cannot be solved legally under any zoning law known to us at present. Practically all attempts at such have proven unconstitutional.
>
> In our studies in Austin we have found that the negroes are present in small numbers, in practically all sections of the city, excepting the area just east of East Avenue [now IH 35] and south of the City Cemetery. This area seems to be all negro population. It is our recommendation that the nearest approach to the solution of the race segregation problem will be the recommendation of this district as the negro district; and that all the facilities and conveniences be provided the negroes in this district as an incentive to draw the negro population to this area. This will eliminate the necessity of duplication of white and black schools, white and black parks, and other duplicate facilities for this area. (Koch and Fowler Engineers 1928) [emphasis added]

There are three conclusions we can reach on the basis of this language. First, its authors, who were consulting engineers to the city, fully recognized that municipal zoning to achieve segregation was unconstitutional, so they adopted an informal spatial classification of the "negro district" that could be enforced not by the police powers that enable zoning but by the allocation of technological infrastructure—sewers, water, electricity, paved streets, street lighting, and schools. If racial minorities wanted such services, they would be provided by the city, but only in the designated "district." In the next few decades following acceptance of the plan, African American citizens of west Austin were denied such service and thus forcibly relocated to the east side of East Avenue, which later became Interstate Highway 35. This part of Austin's story is generally told only on that side of town.

The second point to be made on the basis of the 1928 plan is the ambiguous position of Latinos. There is no reference to "talk about the Latino problem" and, where the plan's plates describe in code African American homes as "miscellaneous residential property," they describe Latino homes as "white residential property." This observation should not suggest that Latinos were considered equal citizens because, where the plan recognized the existence of the traditional Latino neighborhood between the Colorado River and East Fourth Street to be "white residential property," the new zoning ordinance also designated this residential area to be "unrestricted," meaning uses recognized in the courts to be a "nuisance" or what we would now call environmental hazards (Fowler 1928).

A third conclusion to be drawn indirectly from the text is that the existence of racial segregation cannot be attributed to the 1928 plan alone. Rather, the plan was but one means through which social values were materialized (Orum 1987).

Gabriella Gutierez, a veteran east Austin activist, is quick to concur. In her view, it was not only the communities of color that were relocated to east Austin by the 1928 town plan; it was also industry. Zoning of the Latino neighborhoods as "unrestricted industrial" meant that by the 1990s more than 60 percent of the area still permitted industrial use. Thus, the history of space on the east side of town is a checkered zoning policy that randomly mixed minority residential and industrial uses in close proximity to one another.

Minority organizations of resistance did show up in the 1980s, in response to environmental hazards linked to an oil "tank farm," and in the 1990s, in response to a proposal by Sematech Corp. that would have added a major computer chip factory to the east side. Although "high-tech" is understood by many Austinites to mean software programming, a "clean industry," those potentially exposed to the cyanide used in chip manufacturing have a rather different perspective toward its well-documented environmental risks. In developing methods to protect the public health of their own neighborhood, Gutierez and her colleagues have developed a sophisticated understanding of how global economies act on local space with impunity—especially if it is a minority neighborhood. After twenty years of struggle, their experienced voices can now tell the suppressed story of environmental racism in Austin in terms that are not unlike the story of the Frankfurt *Judengasse*.

Latino activists associated with PODER (People Organized in Defense of Earth and her Resources) became effective not only by telling a competing story about their neighborhood but also by backing up that story with empirical data that would have legal standing in the courts. In practical terms this means that they trained high school students to count the number of eighteen-wheelers entering and leaving an industrial site over a twenty-four-hour period and to compare their numbers to the limits established by the city hauling permit. Adult leaders held community workshops on legislative procedures and then sent neighbors to protest at hearings held in the state senate and the county tax district. And they voted. These significant efforts, however, did not become immediately successful or even visible to citizens on the west side.

Lawrence Herson and John Bolland (1998) make the argument that minority rights are least likely to be protected in a city council/manager form of governance because power is so diffuse. Using similar logic, activist Bill Shutkin argues that there is a direct correlation between environmental degradation and the amount of participation and planning on the part of stakeholders in any particular place. In other words, environmental degradation is simply evidence that the local knowledge and interests of some citizens is being ignored (Shutkin 2000). PODER activists provided evidence to support this logic by documenting innumerable incidents when the Austin city manager and city council pointed the finger at each other in decrying conditions in east Austin, yet took no action. Not only has there been a double standard in comparing environmental degradation in east and west Austin, Gutierez argues, but the same has held true for the enforcement of property rights. City planner Clark Allen, who told a valuable part of the story reconstructed in chapter 2, holds that the boom-and-bust economic cycles of the Texas economy tends to recycle the same environmental issues time and time again but in slightly modified form. When pressed a bit further, he acknowledges that there is one constant to the cyclical nature of Austin's environmental talk—that is, the consistent suppression of environmental racism in the city—but where Allen sees Austin's racism as unintended, Gutierez sees it as overt.

In the eyes of east Austinites, Mayor Kirk Watson's Smart Growth Initiative of 1997 was little more than a repeat of the overt racism of the 1928 city plan. In chapter 2, I documented that the Smart Growth Initiative was quickly cobbled together by Watson so that he could appear friendly to the forces of development—who were backed by the state legislature—yet not offend environmentalists. In hindsight, it seems remarkable that Watson never even considered discussing this major initiative with the people who actually lived in the neighborhoods to be most affected. Of course, some will hold that Watson was not inclusive before or after the rollout of the Smart Growth Initiative, but one can still conclude that he considered these citizens as something less than equal. Gutierez takes a rather philosophical approach in saying that

> Smart Growth is a good thing [for the city as a whole], but at whose expense? Declaring east Austin to be the "desired development zone" imposes huge costs on our neighborhoods. Planners have considered land inhabited by people of color to be vacant and available for development . . . and when asked, they even admitted they had not considered the consequences for existing residents.

In all of the interviews I conducted with Austin environmentalists and city administrators not one mentioned the considerable environmental problems of east Austin or mentioned its citizens as collaborators in the environmental movement. In the perspective of Gutierez and her neighbors, however, east Austin residents are in the forefront of environmental protection because they actually use public transit, tend not to consume much water to irrigate lawns, and many in the neighborhood cannot even afford air conditioning, so their per capita rates of electrical power usage are far below average. The irony is that, although east Austinites consume a fraction of the resources than do the romantic environmentalists of west

Austin, they are not perceived as allies in the effort to preserve the environment. To be clear on this point, neither Gutierez nor I argue that all Latinos or African Americans are ideologically "greener" than more affluent whites, only that they lack economic capacity to consume—a reduced economic footprint correlates to a reduced environmental footprint. Of course, the spatial habits of low-income people have been historically linked to many unhealthy practices, particularly regarding food. But if we are interested in the consequences of our habits, rather than attitudes, Rees and Western (2003), among others, have demonstrated that, all things being equal, low income people tread more lightly on the land.

Analysis of environmental controversies in Austin since the 1970s suggests that it is unlikely that the city's famous SOS Ordinance would have passed its public referendum in 1991 without the support of east Austin citizens. Looking in the other direction, however, environmentalists have been inconsistent, at best, in supporting correction of environmental degradation that has accumulated east of Interstate Highway 35. Because middle-class environmentalists rarely travel through those neighborhoods, out of sight is apparently out of mind. This checkered record of support for equity issues by environmentalists has, I will maintain, made it very difficult for a new coalition to emerge that might challenge the traditional dominance of urban policy by real estate interests and rugged individualists. Having been slighted so many times it is hard to imagine that the citizens of east Austin will again support a story told by romantic environmentalists that does not clearly embrace social equity.

Curitiba's Income Distribution

In chapter 3, I argued that in the context of Brazil as a whole, Curitiba has achieved a somewhat narrower or more just distribution of income. Unfortunately, this achievement is a very modest one. When comparing Brazil to its peers in the ten largest national economies in the world, Brazil is at the bottom of the Gini Index, which "measures the extent to which the distribution of income (or consumption) among individuals or households within a country deviates from a perfectly equal distribution" (Bartelby 2003). Although Curitiba has developed a functional middle class, it has not yet succeeded in narrowing the income gap between the very rich and the very poor to levels acceptable to Europeans.

Figure 5.16 illustrates the spatial distribution of income in the city in percentages of the *salario minimum*, or minimum wage, rather than in U.S. dollars. Although these data are not directly commensurable with that for Austin, they do illustrate the spatial distribution of income equally well. By comparing the patterns shown here to those of Austin we can draw some very interesting, if tentative, conclusions about both cities. First, the

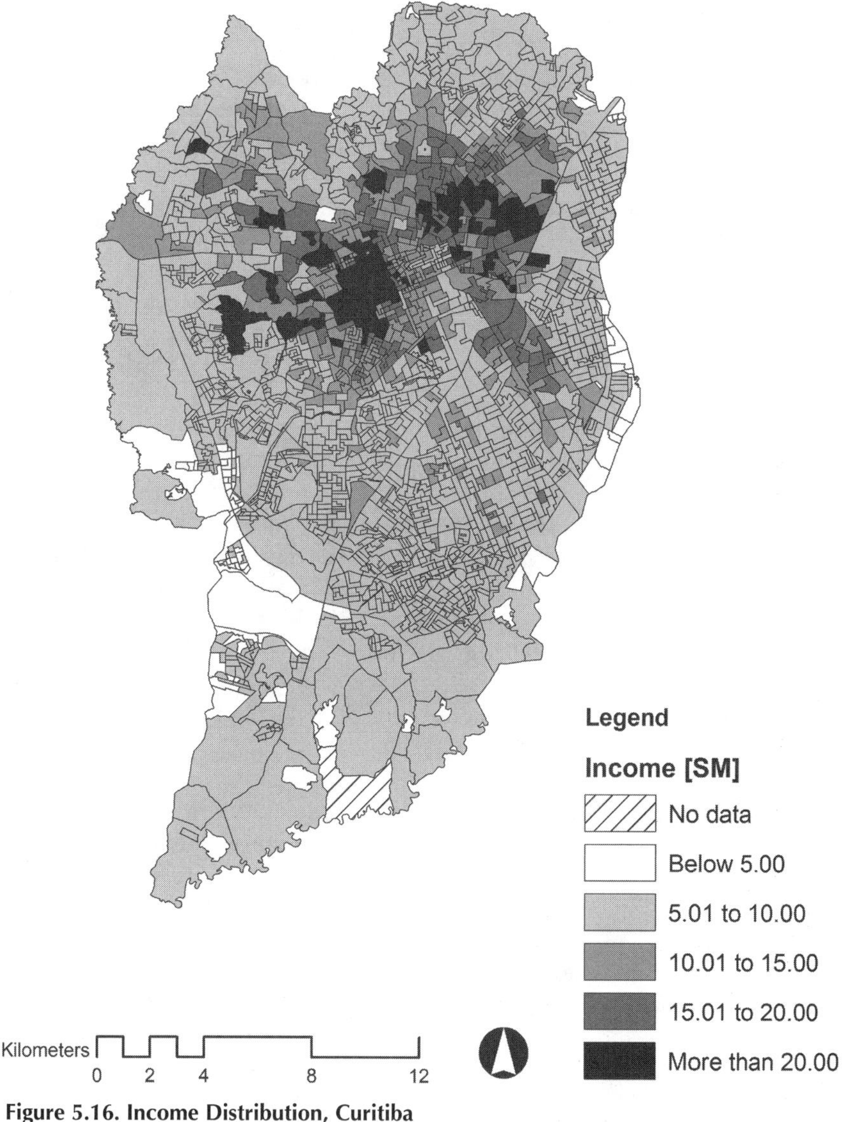

Figure 5.16. Income Distribution, Curitiba

overall pattern in Curitiba is more fragmented than Austin, meaning that the rich, the middle class, and the poor live in closer proximity to each other. Especially in a city with an advanced public transportation system, it means that people from different economic classes are far more likely to encounter each other on a daily basis. The very rich, of course, avoid such contact by driving and living in gated communities, which are increasingly common in Curitiba. Compared to Austin, however, public space is in general more diverse in Curitiba, which I will argue, following Richard Sennett (1998), is a primary criteria for democracy.

A second observation that can be made about Curitiba is that higher incomes are not always correlated to lower densities as has historically been the case in Austin.[5] In fact, the highest densities in the city center are correlated to some of the highest incomes—more than twenty times the minimum wage. This pattern suggests a more urban life style in Curitiba as compared to the generally more suburban lifestyle of Austin.

A third observation to be made is that the highest household incomes in Curitiba, except in the city center, tend to fall between the structural axes that are served most directly by high-speed buses. It should be no surprise that people of means would choose to live in less-dense, less-noisy, and less-polluted spaces. Equally interesting is that the southern structural axes do serve people of modest income. This is particularly true in the most recently constructed arm of the starburst pattern. It is too early to determine whether these areas will eventually be gentrified, but for the moment it is fair to say that people of modest income, and who live in relatively sparse neighborhoods, are far better served by public transportation than their counterparts in Austin. As I argued previously, having access to public space via transit is a primary strategy to overcome the spatial segregation of the poor with all its attendant negative consequences.

The Third "E"

In summarizing the distribution of economic inequality in the United States and Brazil, it is fair to say that one consequence shared by both nations is the production of an urban "underclass" that suffers from what economist James Galbraith refers to as "cumulative inequality" (Galbraith and Berner 2001). This is not to say that the United States and Brazil are the same, only that the economic inequality of some east Austinites and some residents of Curitiba's *favelas* is so persistent and extreme that it precipitates a kind of social isolation that is spatially enforced. Citizens of both countries who live in such conditions tend to develop dead-end or hopeless story lines and norms of behavior that are inconsistent with the development of the social habits that are required to work and participate in the public talk of the wider community (Imbroscio 1997). From a pragmatist point of view, the loss of these stories also has severe consequences for society because it is citizens who live closest to and experience degraded environments who are likely to understand them best. In short, theirs is valuable knowledge that would benefit the city as a whole.

In most public talk related to sustainable development, the question of social equity is usually cast as the weakest of the 3 Es discussed in chapter 1. Although given lip service, few citizens consider it with the same seriousness afforded new and efficient technology or preserving particular natural landscapes. Economist Michael Oden, however, makes a very coherent argument in favor of equity as a necessary, equal, and central component of sustainability.

For Oden, the first problem is to attempt a definition of what we mean by the term *equity*. No one is likely to take seriously the notion that everyone should be equal if it means being exactly the same, but many traditions of thought do argue that all people should be *treated* the same in some particular context: Libertarians argue, for example, that all citizens should have equal property rights, neoclassical economists argue that all citizens should have equal access to the market, and bicycle enthusiasts argue that all citizens should have equal access to city streets no matter what means of mobility they select. Each of these perspectives relies on the understanding that being equal in one privileged context does not translate to full equality because accumulated social capital shapes inequalities in other contexts. Even if access to resources is equal at the beginning of the day, this logic suggests, differences in wisdom, innate ability, or timing will surely make outcomes less than equal by nightfall. It is unlikely, then, that anyone will seriously argue that citizens should be granted full equality in all social contexts.

Oden relies on Michael Walzer's notion of "complex equity" to make the second step toward a proposal. Both agree that it is not likely that we will get citizens of even a small homogenous society, let alone the United States or Brazil, to ever agree on what social context should be most privileged—be it property, markets, streets, or yet some other social context. But rather than throw his hands in the air, Walzer pragmatically proposes that we should simply make sure that inequalities in one social context do not shape or structure inequalities in another. The funding of public schools in Texas is a good example of Walzer's concern because taxes are levied and spent on the basis of real estate value by geographic unit, thus ensuring that the children of the rich will be better educated and accrue greater social capital than those of the poor. This example is only one indicator that in liberal capitalist societies it is considered almost neutral and natural that very successful business people, for example, would also exercise considerable authority in the worlds of art, athletics, education, politics, and even religion. Oden's argument is that complex equity, and a truly sustainable society, would guard against precisely such collateral influence. But why? Because oligarchic authority is unearned and leads to the exclusion of people, voices, and ideas that are necessary to sustain an open, dynamic society.

The third step in Oden's logic is that to flourish, even to sustain itself over time, a society must achieve a degree of "solidarity" among its citizens. By this term Oden means, I think, the particular kind of mutual respect that humans can, but do not always, grant each other. Richard Rorty has put it this way:

> What unites . . . [us] with the rest of the species is not a common language but just a susceptibility to pain and in particular to that special sort of pain which the brutes do not share with the humans—humiliation. (Rorty 1998)

This is to argue that when some citizens are subjected to the humiliation of extreme poverty and social isolation, they can muster neither self-respect nor the respect for others on which civil society, markets, or streets depend to function.

The granting, or denial, of respect to fellow citizens has two implications of a more general nature. First, the exclusion of large majorities from meaningful participation in public decision making breeds a feeling of powerlessness and apathy across the various spheres of life. And second, persistent exclusion of majorities leads to the detachment on the part of elites from the concerns of the citizenry and rigidity in decision making. These two factors, more generally than the denial of respect, undermine the social solidarity and fellow feeling necessary for civic vibrancy, participation, and innovation. Citizens will not participate if they have no impact, and they can have no impact if elites dominate all spheres of social life. The stifling of public participation leads only to path dependence on unsustainable processes and technologies linked to limited isolated interests.

Equity, then, is a relative, not an absolute, condition but relative and complex. Improving equity is not an end in itself, but a process whereby the excluded find reason and ways to band together to shift distributions of power and resources in the various spheres, protecting for example, the political sphere from being dominated and determined by those with inordinate economic power. It is through this kind of public action that respect and social solidarity are formed and decision-making processes in the various spheres of life are opened up to deeper, richer, and more complex information. In this sense movement toward equity is an instrumental "social good" because more diverse and dynamic processes tend to yield more innovative outcomes that benefit all of society, not just the poor.

This discussion of equity and income distribution has generally been rather abstract, so it will be helpful to provide a concrete example that demonstrates its centrality in any talk about environmental sustainability. First, for environmentalists to succeed in Austin, Curitiba, Frankfurt, or elsewhere, land must be removed from the market to preserve the natural conditions or ecological services that are valued. Because there is only so much land, taking some of it out of the market means that the price of the remaining land will necessarily increase to meet even constant demand with reduced supply. This consequence will certainly be received as bad news by those interested in social equity, because increased land cost will be directly translated into higher housing costs. In other words, preserving land can have adverse economic consequences for the poor, especially when they are in close proximity.

But, second, the naturally conflicting interests of preservation and social equity are not insurmountable. Environmentalists might, as Oden argues, support a living wage, greater density (more housing on a given area of land), or affordable housing subsidies so that citizens of low and moderate incomes could afford higher costs and the city could remain diverse. Equity and social justice groups might correspondingly support green building initiatives that would have positive environmental consequences in their neighborhood, in the city, and even globally. The point here is that environmental preservation groups and social justice groups do, in fact, make ideal collaborators in the struggle against the "growth machines" dominated by traditional elites and real estate development interests. They tend, however, to remain isolated from each other because they have different vocabularies, or literally speak different languages, as is often the case in Austin.

Although this chapter has focused on Austin and Curitiba, some readers will by now recall that the success of Frankfurt's sustainability regime was built on just such a "red/green" coalition that linked social equity to environmental preservation interests. The Social Democratic progressives and the Green Party environmentalists were able to forge a common vocabulary with a story line powerful enough to overcome that of the conservative Christian Democrats. In hindsight, it is easy to say that both Austin and Curitiba would benefit from such tactics in the future, but red/green coalitions are built over time by public talk in particular places—they are not forged as short-term backroom deals, as Austin activist Gabriella Gutierez knows only too well. This analysis of space and income distribution does suggest, however, that vital public conversations about sustainable development might begin just as well in those concealed urban spaces such as east Austin or the *favelas* of Curitiba where environmental risks are experienced firsthand rather than in the offices of the big ten environmental organizations. Oden's argument about the relevance of the third "E," equity, simply makes sense (Oden 2005).

But simply arguing that achieving greater social equity will be good for the rich and the poor alike will fail to convince those skeptics who are certain that public participation in decision making will only make marginally talented public servants even more inefficient than they already are. There is, however, empirical evidence that suggests otherwise. Margrethe Winslow, for example, has explored the relationship between environmental quality—a condition highly attractive to "footloose capital"—and democracy by using regression analysis of three persistent air pollutants and two measures of democracy. She found a "robust negative correlation," which means "the higher the level of democracy, the lower the ambient pollution level" (Winslow 2005). Winslow's finding do not prove that democracy "causes" improved air quality, but it certainly does suggest that where more people are directly engaged in making public choices in general air quality is better, and where air quality is better any businessman will tell you that investment is more likely than in locales with bad air quality. Including equity in our operating definition of sustainability makes theoretical sense because many minds are always better than a few. Not only do multiple perspectives see the landscape more thoroughly, but they challenge each other to be more innovative. The proposition also makes empirical sense—both the qualitative and quantitative findings of this study support the purely quantitative analysis by Winslow and others.[6]

5.6 SUMMARY

In concluding this chapter it seems necessary to answer two questions. First, if space conceals some stories, what stories were revealed by using the quantitative tools of geography? And second, how have these stories physically shaped the cities under investigation? To answer these questions I will revisit the categories of analysis introduced earlier.

Reconstructing the evolution of development in the three cities revealed that the quality of the quantitative data available for each city varies significantly and tends to reflect dominant story lines as much as physical reality. Austin's map is deceptive because it literally has holes in it, Curitiba's is deceptive because it is overdetermined by political goals, and on the basis of nothing more objective than my own street-walking experience, Frankfurt's comes closest to reflecting both goals and reality. In spite of these problems it is still reasonable, I think, to hypothesize that different kinds of public talk about politics, the environment, and technology do lead to different patterns of growth on the ground.

Reconstructing the pattern of public parks and open space revealed that local norms can be significantly different. Even if the citizens of Curitiba do believe the story told to the world by the Lerner regime—that their city is a lush landscape—compared to Austin and Frankfurt, it is not. Absolute measures of open space per person are, however, less important than immediate access to some open space for people living in the extremely high densities along Curitiba's axes. In this regard, Curitibanos do not have equitable access to public open space. The technocratic stories told by the Lerner regime have shaped the city to be sure, but this particular story is not lived by citizens. In contrast, the stories told about "the Hill Country" by Austinites and "primeval forests" by Frankfurters has significantly shaped their cities and lives.

The mapping of public transportation in the three cities reveals that Curitiba and Frankfurt live up to their reputations. If we consider transportation planning to be a kind of storytelling, the planners of Frankfurt and Curitiba have convincingly narrated alternative futures that citizens have embraced. Not so in Austin. In spite of the attractive stories told by Cap Metro—the regional transportation authority—the majority of citizens, including the minority communities of the east side, have not embraced them as their own.

Mapping the relative densities of our three cities reveals Austin to be a rather sparse and suburban city compared to Frankfurt and to Curitiba, in particular. It seems that the rugged individualists of Austin are more responsive to stories about lonely cowboys than they are to stories lauding the benefits of increased density told by planners, but in this analysis there was evidence that citizens could talk themselves out of their natural aversion to density, but only when they could describe such space in their own terms.

The mapping of income distribution in two cities revealed Austin to be an economically and racially segregated city and Curitiba to be more like other South American cities than the stories told by the Lerner regime would have it. Yet social equity, no matter how relative or hard to define, is an essential characteristic of sustainability for entirely instrumental reasons—it contributes to the solidarity and creativity on which societies, markets, and ecosystems depend to regenerate themselves successfully. Frankfurt proved to be the only case in which a coalition of the poor and environmentalists—the so-called red/green coalition—managed to articulate and achieve solidarity among a majority of citizens for a significant period of time.

When considered as a group, the spatial patterns reconstructed in these three cities tend to pair up in surprising ways. Regarding evolution of development and open space, Austin and Frankfurt are superficially similar, but Curitiba is quite different. Regarding public transit and density, Curitiba and Frankfurt are quite similar, but Austin is quite different. And regarding income distribution, the data are incomplete, but Austin and Curitiba are more similar than one might expect.

The thick stories of these three cities do not add up to anything that I would attempt to call a metaspatial pattern—there are too few samples here to support such a claim. The evidence does suggest, however, that there is considerable diversity in the spatial patterns in these cities that aspire to develop sustainably. And because the history of space in each city is so different it is not likely that their trajectory of development will become more similar over time. Austin, for example, is certainly becoming denser, but is not likely to ever approach the densities achieved in Frankfurt, let alone Curitiba. Rather, the city will seek out those paths of sustainable development that are more consistent with the spatial habits and stories told by its citizens.[7] The problem will, of course, be how to make new spatial habits, like racial integration, attractive.

Finally, I do not want to leave the reader with the impression that local spaces and story lines are constructed in isolation from the structural forces and constraints associated with globalization. Although I have not stressed the local impact of international institutions like the World Bank, there are few spaces that manage to avoid their gravitational pull. But in the end all spaces are lived locally.

CHAPTER REFERENCES

Agyeman, J., R. Bullard, and B. Evans, eds. (2003). *Just Sustainabilities: Development in an Unequal World*. London: Earthscan and Cambridge, MA: MIT Press.

Baker, L. A., A. Brazel, N. Selover, C. Martin, N. McIntyre, F. R. Steiner, A. Nelson, and L. Musacchio. (2001). "Local warming: Feedback from the urban heat island." Unpublished ms, 25 October 2001. Collection of the author, University of Texas.

Bartelby. (2003). *Distribution of family income: gini index*. <www.bartleby.com/151/> (accessed 2 January 2006).

Beck, Ulrich. (1992). *Risk society: Towards a new modernity*. Thousand Oaks, CA: Sage.

Berger, J. (Writer) (1974). *Ways of seeing*/BBC-tv. In M. Dibb, producer. UK: Films Incorporated; Wilmette, IL.

Burton, E. I., ed. (2000). *The potential for the compact city for promoting social equity*. London: Spon.

Cap Metro. (2005). Advisory committee minutes for August 17. <www.capmetro.org/news/publicmeetings.asp> (accessed 1 September 2005).

Chadwick, E. (1965). *Sanitary conditions of the labouring population of Great Britain*. Edinburgh: Edinburgh University.

Foucault, M. (1977). *Discipline and punish: The birth of the prison*. New York, Pantheon.

Galbraith, J. K. and M. Berner (2001). *Inequality and industrial change: A global view*. New York: Cambridge University Press.

Gregory, D. and J. Urry, (1985). *Social relations and spatial structures*. Basingstoke, UK: MacMillan.

Guy, S. and S. A. Moore. (2005). *Sustainable architectures: Natures and cultures in Europe and North America*. London: Routledge/Spon.

Hall, P. G. (1996). *Cities of tomorrow: An intellectual history of city planning in the twentieth century*, 2nd ed. Oxford: Blackwell.

Harvey, D. (2000). *Spaces of hope*. Berkeley, University of California Press.

Herson, J. R. and Bolland, L. (1998). *The urban web: Politics, policy and theory*, 2nd ed. Chicago: Nelson-Hall.

Hillier, B. and J. Hanson. (1984). *The social logic of space*. Cambridge: Cambridge University Press.

Imbroscio, D. L. (1997). *Reconstructing city politics: Alternative economic development and urban regimes*. Thousand Oaks, CA: Sage Publications.

IPPUC. (2005). *Instituto de pesquisa e planajamento urbano de curitiba*. <www.ippuc.org.br/>. (accessed 10 March 2005).

Irazabál, C. (2005). *City making and urban governance in the Americas: Curitiba and Portland*. Hants, England: Ashgate.

Kassen- und Steueramt, Frankfurt am Main. (2005). Private acquisition of data from transaction ID: 1-12/0-4011-Ktr. 00920 Bel.: 19.

Koch and Fowler Engineers. (1928). *Plan for the city of Austin*. Austin, TX: Austin City Plan Commission.

Lefebvre, H. (1991). *The production of space*. Translated by Donald Nicholson-Smith. Cambridge, MA: Blackwell.

Moore, S. A. (2001). *Technology and place: Sustainable architecture and the blueprint farm*. Austin: University of Texas Press.

Morello-Frosch, R. (1997). "Environmental justice and California's 'riskscape': The distribution of air toxics and associated cancer and non cancer risks among diverse communities," unpublished dissertation, Department of Health Sciences, University of California, Berkeley.

Newton, P. (2000). "Urban form and environmental performance." Pp. 46–53 in K. Williams, E. Burton, and M. Jenks, eds., *Achieving sustainable urban form*. London: Spon.

Oden, M. (2005). "*The question of equity*." Lecture delivered April 10 at the University of Texas, School of Architecture.

Orum, A. (1987). *Power, money, and people: The making of modern Austin, Texas*. Austin: Texas Monthly Press.

Pinderhughes, R. (2004). *Alternative urban futures: Planning for sustainable development in cities throughout the world*. Lanham, MD: Rowman & Littlefield.

Rees, W. and L. Westra. (2003). "When consumption does violence: Can there be sustainability and environmental justice in a resource-limited world?" Pp. 99–124 in J. Agyeman, R. D. Bullard, and B. Evans, eds., *Just sustainabilities: Development in an unequal world*. Cambridge, MA: MIT Press.

Rorty, R. (1998). *Achieving our country*. Cambridge, MA: Harvard University Press.

Sennett, R. (1998). "The spaces of democracy." Paper presented as the Raoul Wallenberg Lecture, University of Michigan, College of Architecture & Urban Planning. Ann Arbor, MI.

Shutkin, W. (2000). *The land that could be: Environmentalism and democracy in the twenty-first century*. Cambridge, MA: MIT Press.

Soja, E. W. (1989). *Postmodern geographies: The reassertion of space in critical social theory*. London/New York: Verso.

Taylor, I., K. Evans, and P. Fraser. (1996). *A tale of two cities: A study in Manchester and Sheffield: Global change, local feeling and everyday life in the north of England*. London: Routledge.

Torras, M. and J. K. Boyce (1998). "Income, inequality, and pollution: a reassessment of the environmental Kuznet's Curve." In *Ecological ecolomics* (25): 147–60.

Williams, K. E. B. and Mike Jenks, eds. (2000). *Achieving sustainable urban form*. New York, Spon.

Winslow, M. (2005). "Is democracy good for the environment?" *Journal of Environmental Planning and Management*, 48 (5): 771–83.

NOTES

1. From a post-Puritan American perspective I should also note that the nineteenth-century religious reform movements associated with the Second Great Awakening in North America softened the harsh sentence imposed upon the poor by both American Puritans and earlier European Christians.

2. Edwin Chadwick, whose close association with the Utilitarian philosopher Jeremy Bentham gave him considerable public visibility, is generally credited with being the first to argue that the ill health of the poor was caused by the degraded environmental conditions in which they lived rather than moral depravity. Ian Taylor has, however, documented that Kay-Shuttleworth issued a report based upon similar logic in 1832 on the conditions of the poor in Manchester that predates Chadwick's report to Parliament of 1842. The point here is not to challenge Chadwick's contribution to the modern public health movement but to place his influence within the public talk of that era.

3. This argument is not made with reference to "informal" settlements, which may not be related to either planned development or developer profits.

4. These quantities are from three different sources using different data and should be considered very approximate. I have used reported total annual single trips divided by population within city jurisdictional boundaries for comparison.

5. After 2000 this pattern began to change in Austin with the construction of several high-rise "lofts" for upper-income citizens.

6. These findings are amplified by the work of Julian Agyeman, Robert D. Bullard, and Bob Evans (2003) who document other correlations between social equity and environmental quality. Torras and Boyce (1998), for example, have shown that "in countries with a more equal income distribution, greater civil liberties and political rights, and higher literacy levels tend to have higher environmental quality." Likewise, a study by Morello-Frosch (1997) "of counties in California showed that highly segregated counties, in terms of income, class, and race, had higher levels of hazardous air pollutants." In general, it is fair to say that there is mounting evidence that correlates democratic processes, in the sense intended here, to environmental quality and economic development.

7. In 2005 the City of Austin announced a plan developed by Austin Energy, the city-owned electrical utility, to support the manufacture and sale of electric hybrid cars. The economic incentive for the plan derives from the fact that electrical utilities must overproduce capacity between peak consumption periods. By selling off-peak power to the owners of electric cars at greatly reduced rates the city would not only gain income but reduce auto emissions and improved air quality. This very inventive plan is highly compatible with the values of Austinites, which emphasize mobility and suburban open space.

SIX

Sustainability and Democracy

This chapter has two purposes. It begins with a summary of the kinds of story lines, public talk, and spatial organization found in each of our three cities. This exercise is designed to emphasize any similarities or differences that may exist. The larger part of the chapter then considers twelve dilemmas that emerge from my analysis of those similarities and differences.

On the basis of table 6.1, I hold that some kinds of public talk were shared in the three cities, but not nearly enough to argue that any particular kind of talk is a necessary condition for sustainable development to show up. As I have argued earlier, the political dispositions of these cities are very different. Regarding environmental talk, however, there is some overlap. Five of the environmental discourses identified by John Dryzek show up in the three cities: "green romanticism" is a kind of talk shared by Austin and Frankfurt, and "green rationalism" is a kind shared by Curitiba and Frankfurt. This amount of overlap is, however, not surprising in any conversation about environmental sustainability. As I argued in chapter 5, the evolution of development of Austin and Frankfurt is superficially similar, even if the leapfrog pattern appeared for different reasons. The same can be said for the distribution of parks and open space, which is similar in Austin and Frankfurt but not in Curitiba. When the data for public transportation and density are combined, there is a pattern that shows Austin to be the least sustainable and Curitiba to be the most if we consider transit-oriented development to be a primary criterion. More significant, however, is the quantitative spread between cities that are already positively associated with sustainable urban development. Finally, the income distribution data, which were not available for Frankfurt, show that Austin is rigidly segregated and that Curitiba has segregated islands but far more edges where contact occurs.

In all, the kinds of stories told and the spatial characteristics developed in these cities demonstrate no consistent pattern that we can associate with the appearance of sustainable development. Articulating this finding

Table 6.1. A Comparison of Narrative and Spatial Characteristics

	Austin	*Curitiba*	*Frankfurt*
Story lines	Rugged individualism versus environmental preservation	Technocracy versus social democracy	Progressive capitalism versus red/green
Political talk	Liberal anarchism versus progressive populism	Liberal realism versus citizen participation	Liberal minimalism versus strong democracy
Environmental talk	Economic rationalism versus green romanticism	Administrative rationalism versus green rationalism	Ecological modernization versus green romanticism and green rationalism
Technological talk	Self-creation versus clean technology	Technological display versus ad hoc	Soft determinism versus soft voluntarism
Evolution of development pattern	Leapfrog	Concentric	Leapfrog
Parks and open space in square meters per person (ratio)	68.39 (7.3)	9.32 (1.0)	97.99 (10.5)
Public transit in citizen trips/year (ratio)	52 (1.0)	456 (8.8)	336 (6.5)
Density in persons/hectare (ratio)	16.9 (1.0)	102.5 (6.0)	63.2 (3.7)
Income distribution	Segregated east and west	Segregated islands	No data

in reverse is to say that there is more than one path to the sustainable city. One size does not fit all. But this finding is not to say that nothing more can be learned from the investigation other than challenging the notion that all cities must develop in accordance with a single model of sustainable development. To the contrary, the lessons learned are that it is not only the plot of local stories that is important but also the conflicts or social dilemmas that arise from their telling.

In the way I employ the term here, a dilemma is a situation in which one is presented with two or more alternative choices that seem equally reasonable or unreasonable—at least as abstract propositions. As I have argued in previous chapters, however, the apparent equality of the choices is generally changed when one has a situated perspective of the conditions at hand. When viewed through the local history of space the relative attractiveness of alternative routes is altered and may not be the same for any two cities when facing the same dilemma—history and space matter.

The order of presenting these dilemmas has a loose rationale but might easily be revised, expanded, or contracted. Likewise, the generally dualistic structure of the dilemmas themselves is more for rhetorical purpose than it is substantive. The reader will soon recognize that the symmetry of good/bad or light/dark is only of rhetorical value. My intent in structuring the analysis in this way is to allow the theory developed in the previous chapters to be tested by the empirical case studies, just as the cases are tested by the developing theory. The analysis that follows, then, is neither comprehensive nor easily generalized.

The question of comprehensiveness is not a difficult one to settle. In chapter 1, I forewarned readers that the stories of Austin, Curitiba, and Frankfurt were "thick," filled with many actors, plots, and subplots. The twelve dilemmas that I have gleaned from these stories are simply those conflicted themes that I found to be most important to the questions posed at the outset. Other dilemmas and observations can surely be constructed from the collected data, but there comes a point when redundancy and reader fatigue sets in.

The question of generalizability is, however, more difficult. In previous chapters I argued that the type of qualitative and quantitative analysis conducted in this study cannot simply be transported from one situation to another—people do not behave with the same degree of logical consistency as do atomic particles. Such uncertainty, however, does not mean that analysis is useless. Rather, it means that it becomes the responsibility of citizens or subsequent analysts to determine the appropriateness of the "fit" between the dilemmas encountered in these cities and those encountered in others in the future.[1] This is to say that the purpose of constructing these dilemmas is to begin new conversations, not to preserve old ones.

6.1 STORIES AND SPACES OF COALITION

In the previous chapter I tested how local stories corresponded to the condition of urban spaces. I found that the lenses of history and those of geography do not always agree. Nevertheless, this finding does not suggest that we should privilege geography over history, but that both tools are needed to reconstruct reality as best we can.

The sustainable city may, in the end, prove to be a utopian project. If we are to make any progress at all, however, this analysis demonstrates that it will be necessary for us to design the spatial forms preferred by geographers along with the social processes preferred by historians (Harvey 2000). Another way to say this is that urban form is related to how public talk describes the city just as the content of public talk is related to the history of a particular space. Or better, spaces create publics and publics create spaces.

A corollary to this logic is that strongly democratic publics are constituted of coalitions and that coalition building gets harder when constituencies are separated by spatial and linguistic barriers. When a substantial portion of the citizenry is inaccessible to you, it is far more difficult to understand what you may have in common and how you might act together.

The lesson to be learned from the cities studied is that more sustainable systems are created not only by rigidly adhering to principles (as in Austin) but by building coalitions (as in Frankfurt). Successful activists in all three cities have not harangued fellow citizens with an alien vocabulary of "alternative," "green," "regenerative," or "sustainable" choices, but they have redescribed the dominant story line and eroded barriers to coalition building by using vocabularies that are historically part of the local public talk.

6.2 TECHNICAL AND CULTURAL RATIONALITY

In each of the cities studied here there have been frequent and volatile public debates concerning deteriorating air quality, threats to water quality, climate change, and numerous other issues. The data provided by scientists that document changing conditions were rarely in dispute. No one in Austin, for example, has contested the presence or quantity of benzene-derivative chemicals in the Barton Springs watershed. The interpretation of the significance of that data for human or ecological health is, however, still bitterly disputed by scientists and laypersons alike (Dryzek 1997).

My analysis of the cases suggests that a partial source for such bitterness may be the distinction between what we might call "technological rationality" and "cultural rationality" (Fischer 2000). Technological rationality is the abstract or symbolic logic associated with the Enlightenment and scientific method. It is this style of thinking that we tend to privilege because we have historically accepted it as more objective and free from those social, political, or economic interests that might sway or distort "the Truth." It is this type of reasoning that led Austinites to adopt a city manager form of government, Curitibanos to seek managerial efficiency, and Frankfurt's bankers to compete for the lowest energy-consumption rates.

On the other hand, we recognize the existence of cultural rationality in conflicting ways. Scientists are apt to dismiss cultural logic as emotional or irrational. William Rees, for example, understands cultural rationality as a type of filter that prevents citizens from being able to assess the alarming environmental data made available to them by science (Rees 2004a). From the perspective of many experts, cultural rationality is an obstacle to both understanding the situation as it really is and conclusive action based on good science.

Successful activists in each of our three cities see the situation quite differently. Unlike Rees they intuitively understand that these two forms of rationality are necessary complements to one another. Each type of reasoning produces a different type of knowledge, and both are required if our concern is to construct a satisfying lifeworld, and not just an efficient one. Thus, cultural logic produces a kind of knowledge that assesses risk in a "situated" rather than abstract manner (Haraway 1995). For example, when scientists calculated the probability of polluting the air of Frankfurt (in particulate parts per million) correlated against the economic cost of asthma attacks, citizens with asthma had a very different standard of what constitutes an acceptable risk. Another way to argue this point is to suggest that "probability" is a scientific term and that "risk" is a cultural term—they do not measure the same thing. To the extent that science is value neutral, it cannot recognize a risk, because it cannot distinguish between good and bad outcomes.[2] This observation should not suggest, however, that "risk" is not real. Successful activists, then, recognize that technological and cultural rationality must be accountable to each other if we are to act successfully.

Scientists are members of an "epistemic community" that is reproduced largely in isolation from the confusing diversity of local practices (Guy 2000). As a result of their isolation, scientists generally fail to understand why their proposals for water purification or energy efficiency fail to attract a majority of local users. Even worse, they imagine that practitioner engineers and citizen activists disagree with their assessments only because they lack access to the latest knowledge or because they are impeded by some nontechnical barrier that lies outside the realm of science. It would simply never occur to some of the scientists involved in Austin's water-quality wars, Curitiba's resettlement protests, or Frankfurt's housing struggle that citizens of other epistemic communities might have a rational and valid perspective of the situation.

Scientists and activists are, however, likely to agree that the gap between technological and cultural rationality has increased as our technologies have become more complex and opaque. Fischer argues that "affected citizens begin to recognize not only that scientists are ignorant of the consequences of their actions but also that scientists are interested laypersons in their own scientific projects" (Fischer 2000). As a result, citizens have begun to equate technological rationality with cultural irrationality and generally reject the authority of science in favor of other value systems. One consequence of this divide is the not-in-my-backyard, or NIMBY, phenomenon, in which citizens reject public policies on the basis of scientific findings because they are deemed both suspect and irrelevant to daily life. The counter-reaction is for scientists, engineers, and the politicians who hire them is to consider citizens as increasingly irrational and self-interested, as was the situation in both Frankfurt and Curitiba during the 1970s and Austin in the 1980s.

One response to this downward spiral in public trust is to replace the symbolic logic of the scientist in urban planning with the practical logic of the judge and the planner. Judicial and planning choices—at least the most successful ones—are made in the messy context of daily life rather than in the reductive context of the laboratory. As I noted in chapter 1, Charles Peirce, following Aristotle, referred to this type of logic as "abductive," and in a modern context we might refer to this mode of rationality more simply as *critical thinking*. For the sake of historical continuity, however, I will employ the term *abduction*.[3] Successful judges and planners tend to be skilled at this form of hypothesis generation because they are practiced at reading complex human situations, not because they have an infallible method.

I argued in chapter 3 that the Lerner regime employed abductive reasoning in the planning of infrastructure in Curitiba—the same can be said for the planning of a new type of tall building in Frankfurt or the development of Austin's watershed zoning program. In all cases planners and architects were able to invent new and successful urban typologies because they employed local knowledge to solve universal problems rather than extruding local spaces through universal principles. Perhaps most fundamental to abductive reasoning is that it is not blind to human values but situated within the community looking toward the future rather than at the past.

In sum, I must emphasize that my proposal to employ abductive reasoning in city making should not be construed as an attack on the legitimacy of traditional science. Rather, my argument is simply that both forms of reasoning are required to build a satisfying and still efficient lifeworld.

6.3 EFFICIENT AND INCLUSIVE SYSTEMS

> Strong democracy "tries to revitalize citizenship without neglecting the problem of efficient government by defining democracy as a form of government in which all of the people govern themselves in at least some public matters at least some of the time."
>
> —Benjamin Barber, *Strong Democracy*

In the development of political or technological systems we, like the mayors of Austin and Curitiba, tend to think that including all of the people affected by decision making would be terribly inefficient. From personal experience we all can summon up the rolling eyes of coworkers when someone suggests yet another committee meeting. Likewise, we malign the aesthetic value of camels by asserting that their origin suffers from "design by committee." The implication of such passive dissent is that group decision making is a waste of valuable time that might be more efficiently spent in getting stuff done.

Those who favor representative forms of government, for example, generally do so on grounds of efficiency. The advocates of representation inevitably provide examples of how mass movements, like the ones that surfaced in Frankfurt in the 1960s or in Austin in 1990, degenerate into the proliferation of individual interests or simply die out altogether. Likewise, the advocates of both enlightened dictatorship and technocracy, as in Curitiba, make the same argument but have an even more skeptical view of the value of participatory democracy.

In the world imagined by technological rationalists such as the members of the Lerner regime in Curitiba, efficiency is assumed to be a public "good." It follows that inefficiency—the unconscious waste of resources—is a public "bad." This set of values has been a part of Western culture for some time and is generally attributed to the rise of utilitarianism in Britain during the nineteenth century. Jeremy Bentham (1748–1832) and his younger associate, Edwin Chadwick (1800–1890), were principals in the development not only of utilitarian philosophy but also of public works projects based on the doctrines of what they referred to as "civic economy" (Chadwick 1965). In 1842, Chadwick reasoned that the ill health of English workers was not only a source of misery to them but also a "pecuniary burden" to the economy as a whole (Chadwick 1965, 253). In short, utilitarians were the first to argue that poor public health was inefficient. The project of civic economy was for an exclusive group of "civil engineers" to design new forms of political and technological systems that would efficiently exercise the power of the state in a manner that would result in "morals reformed—health preserved—industry reinvigorated—[and] public burdens lightened" (cited in Foucault 1977, 206).

As I briefly noted in chapter 5, Michel Foucault has famously criticized the technocratic manner in which the utilitarians exercised power by linking the institutions of government to environmental surveillance of various kinds (Foucault 1977). In his assessment, the efficient proposals for prisons, hospitals, and infrastructure of all kinds made by utilitarians were designs of "subtle coercion" (Foucault 1977, 209). Foucault's critique aside, the utilitarian assumption that there is an inherent conflict between the design of efficient and inclusive systems is simply too sweeping and inconsistent with my analysis of our three cities—Frankfurt in particular. Whether it is in the design of institutions or the buildings that house them, "design is not a zero-sum economic game but an ambivalent cultural process that serves a multiplicity of values and social groups without necessarily sacrificing efficiency" (Feenberg 2002, 95). Science and technology studies scholars are fond of arguing that technology is underdetermined by the criteria of efficiency. This is to argue that, even after all of the needs of economy are met, there is still significant interpretive flexibility in design that the system might be realized in a myriad of ways. Bentham, Chadwick, and Jaime Lerner would, no doubt, argue that the political and economic turmoil of their respective times required a strong central authority to improve the lot of proletarians in spite of themselves. Perhaps. But this claim does not provide any evidence to thwart the counter-claim that an inclusive design process would be more *effective*.

The fixation on efficiency, promoted in Curitiba by technocrats and in Frankfurt by the supporters of ecological modernization, limits the goals of production to what is scientifically possible. The argument here is that by setting our goals on what is socially desirable, rather than on what is scientifically possible, our choices will, in the end, be more effective because they are understood at the outset as sociotechnical in scope. Although time is surely expended in the process, the inclusion of multiple perspectives in the design of artifacts and institutions renders them more satisfying.

Frankfurt's twenty-year struggle to articulate an acceptable planning scheme for its banking district is certainly the best example of planning in the cases studied. From inside that struggle during the 1970s and 1980s the endless debates must have seemed an enormous waste of human resources to some. City planner Helmut Bosch, however, now sees the effort as a very sound investment of time by citizens, government, labor, and business because it produced an environment that works. Just as the builders of the Commerzbank tower invested time, labor, and money to test construction methods on a full-scale prototype of a single floor of the tower before actual construction got underway, planners and citizens invested years in the evaluation of alternative plans of the district as a whole. In the long history of this city, the planning process is now viewed as inclusive, efficient, and effective because it is working—it is being sustained.

6.4 SCARCE AND ABUNDANT RESOURCES

The debate between those who perceive natural resources to be scarce and those who perceive them to be abundant is at least as old as Thomas Malthus (1766–1834) and Karl Marx (1818–1883). This debate has continued, mostly unchanged, among the citizens of Austin regarding water, in Curitiba regarding forests, and in Frankfurt regarding natural resources in general.

Malthus's hypothesis is "that population, when unchecked, increased in a geometrical ratio, and subsistence for man in an arithmetical ratio." This assertion was initially stated in his *Essay on the Principle of Population* of 1798 and from it he further reasoned:

> But though the rich by unfair combinations contribute frequently to prolong a season of distress among the poor, yet no possible form of society could prevent the almost constant action of misery upon a great part of mankind, if in a state of inequality, and upon all, if all were equal.
>
> The theory on which the truth of this position depends appears to me so extremely clear that I feel at a loss to conjecture what part of it can be denied. (Malthus 1798)

Translated into more contemporary language, Malthus reasoned that although the rich do distress the situation of the poor, no form of governance could overcome the fundamental tension between population growth and resource scarcity that is the true source of human misery. Even worse, he argues, any attempt to achieve social equality would only exasperate misery and social ill temperance. This logic is employed to rationalize the natural dominance of the rich and their control of population growth through warfare, prostitution, and managed disease so as to save both civilization and nature from decline and inevitable collapse. Malthus was nothing if not clear.

What is of interest here is that updated versions of Malthusian logic appeared in all three cities studied. In Austin, rugged individualists evoked Malthusian logic to argue for the natural dominance of the rich, and green romantics evoked it to argue for strong centralized environmental controls. In Curitiba, the Lerner regime of sustainability openly used Malthusian environmental logic to justify a strong centralized government. And in Frankfurt, the *Fundis* of the Green Party employed a form of Malthusian logic that some scholars relate to latent Nazism (Bramwell 1985). In all three cities locals were able to support their views by referring to contemporary neo-Malthusians Malthusians (AtKisson 1999; Bahro 1994; Ehrlich 1968; Hardin 1968; Meadows 1995). Perhaps the best spokesman for the neo-Malthusian position is Rees, an admirer of Curitiba's planning policies who refers to his position as "enlightened Malthusianism"(Rees 2004b). He argues:

> Unsustainability is an old problem—human societies have collapsed with disturbing regularity throughout history. I argue that a genetic predisposition for unsustainability is encoded in certain human physiological, social, and behavioral traits that once conferred survival value but are now maladaptive. (Rees 2004a)

Rees is clearly not optimistic about the ability of Western societies to modify their behavior. As a result, he argues for enlightened "mutual coercion"—a strategy that is certainly less callous than the proposals of Malthus, yet necessary, he argues, to avoid ecological collapse. Neo- (or enlightened) Malthusians, then, accept the unfortunate necessity to suppress human consumption by undemocratic means because if left to our biological habits collapse is inevitable.

From a contemporary perspective, many consider Malthus's assessment to be radical and extreme. We have, after all, doubled and redoubled populations and still managed to generally increase the quality of human life. It is therefore not surprising that Karl Marx and those who follow him found Malthusian logic not only lacking but also politically self-serving. For Marxists any initial scarcity of resources can be altered by the use of human rationality on nature through technology. Marxism, then, turns Malthusian theory upside down by finding abundance, not in resources but in human (mental) labor (Marx 1908).

But just as neo-Malthusians have softened the grimness of their mentor's view of history, neo-Marxists have softened the optimism of theirs. Looking back at Marx's texts from a contemporary vantage point, some hold that toward the end of his life Marx did come to recognize that resource scarcity was not merely a hoax perpetrated by the rich. Rather than lend support to Malthus, however, he buried his observations in his critique of capitalism's inability to use resources efficiently (Benton 1996; Perleman 1996). Thus, the neo-Marxist position sounds rather like the argument for ecological modernization made by Frankfurt's environmentalists and bankers—that we can feed the masses by slicing the bologna ever thinner.

And so, which way shall we have it—are resources scarce or abundant? From a contemporary perspective—that of decision makers in our three cases—it is no longer rational to argue that scarcity is a myth fabricated by the rich to justify suppression of the poor, but it is no more rational to argue with the neo-Malthusians that coercion is the only way we can halt overconsumption of increasingly scarce resources. I make this middle-of-the-road argument on three grounds.

First, coercion has not worked in the past and there is little reason why we should think it will now.

Second, overconsumption is less an ecological necessity than a bad habit enabled by political conditions. Consuming less can be made more attractive by design—an argument that has been supported by empirical evidence (Brand 2005; Rohracher 1999).

And third, both traditional Marxists and Malthusians succumbed to the attractive delusion that history was on their side—that they alone understood the trajectory of human nature. Human nature, and thus history, however, has turned out to be far more complex and unpredictable than the teleology sketched by either side. On the basis of history since Malthus and Marx, it now appears more rational to argue that human nature is developmental and that history is contingent. A discussion of the coevolution hypothesis should make this claim clearer.

6.5 COEVOLUTION OF NATURE AND TECHNOLOGY

Malthusians, as I have already suggested, tend to see human nature as unchanging. Theirs is an elitist view of human kind that owes some of its logic to an aristocratic society into which rationality and agency are thought to be a function of breeding. It is also a view very much at home in contemporary hierarchical societies such as the United States and Brazil, where there is a monumental gap between the conditions enjoyed by the rich and those suffered by the poor. Here I will be particularly critical of the neo-Malthusian position, not only because of its implicit elitism but also because Malthusians incorrectly characterize populations as without agency. Recent history—as in the elimination of apartheid from South Africa, for example—has surely demonstrated otherwise (Dryzek 1997).

Nevertheless, it is important to establish that contemporary liberal capitalists—like the ones who play significant roles in the three cases studied above—"also assume that human nature is essentially unchangeable" (Prugh 2000, 104). But where neo-Malthusians are basically pessimistic about the consequences of unchanging human biological habits, liberal capitalists are basically optimistic. The optimism of liberal capitalists, however, is not focused on unchanging human habits but on the consequences of unchanging human self-interest. To be clear, liberal capitalism privileges greed, not intelligence, as the constant human trait. In its most radical libertarian form, as plays a large part in the story told by Austin's rugged individualists, human self-interest is seen as the salvation of nature because only private property interests can avoid the "tragedy of the commons" (Hardin 1968; Prugh 2000).

In both assessments—for better in the eyes of liberal capitalist or for worse in the eyes of neo-Malthusians—human nature is fixed. The optimism of liberal capitalists, however, is dulled when confronted with the apparent irrationality of markets. Likewise, neo-Malthusian arguments for coercion tend to dull when confronted with the failure of authoritarian societies such as those of Easter Island, the Maya, or countless others. Still, liberal capitalists and Malthusians are certainly not the first or only observers to abandon faith in human rationality and the malleability of the human character. In chapter 7, I will review the general attack on rationality embraced by romantics and postmoderns that emerged in the wake of nineteenth-century industrialization and twentieth-century wars. For the moment, however, I will simply state that it was this growing Western skepticism toward human rationality and urban development that influenced some environmentalists (green romantics) in Austin and Frankfurt. In this context the "hippy" era of the 1960s and 1970s can be understood as anti-Enlightenment and deeply skeptical of technological rationality as I presented it previously.

Here, however, I must distinguish between Enlightenment *rationality* and human *intelligence*. By the former I mean a particular way of reasoning that developed in the eighteenth century, and by the latter I mean the more fundamental ability to reason. This is a distinction that many will find objectionable, but one that is, I think, compatible with pragmatism. John Dewey, unlike most critical theorists of his time and later postmoderns, was generally optimistic about the potential of human intelligence if not of technological rationality. He argued, "If human problems are to be solved it will be human intelligence that will have to do the job. His story is not the myth of hope, or even of salvation, but of human responsibility" (Hickman 2001, 155). Unlike that authored by Malthusians and liberal capitalists, the story Dewey put forward was not based on the notion that human nature was fixed, but the opposite, that human nature is changeable, or developmental.

In Dewey's assessment, thinking is biological in its origins:

> Of human organisms it is especially true that activities carried on for satisfying needs so change the environment that new needs arise which demand still further change in the activities of organisms by which they are satisfied; and so on in a potentially endless change. (Dewey 1991, 135)

Dewey's position is not a form of environmental determinism, but a *relational* understanding of humans and nature that is entirely consistent with contemporary ecology. At its core is an understanding that humans interact with nature through technology. As human projects transform what we can call "first nature" into "second nature," not only are new technologies required, but humans also change in response to altered environmental conditions. In this schema humans, nature, and technology are each granted shifting degrees of agency ad infinitum (Dewey 1991).

I will take care not to leave the impression that human reasoning can direct evolution. Such logic would inevitably lead to a proposal that Frankfurt, for example, has evolved to a greater degree than Austin or Curitiba. Any such claim would require us, first, to know the direction of history from our position within it. To avoid such a simplistic view of history or evolution we should distinguish between three models of evolution: those of Herbert Spencer, Jean-Baptiste Lamarck, and Charles Darwin. Spencer's model proposes that greater complexity and progress are inevitable—his is a type of divine teleology not dissimilar from Marx's. Lamarck's model proposes that human striving drives change but that progress per se is never guaranteed. Darwin's model, however, proposes that evolution requires neither a historical force nor a direction of change. I will quickly dispense with Spencer's model of inevitable progress as wishful thinking, even if it is an attractive story still championed by many. Lamarck's model, however, seems closest to Dewey's position, which is something of a problem because most contemporary evolutionary historians have discarded the idea that the striving of individuals can influence the long course of evolution. In spite of our ability to reason, skeptics argue, we can never predict the precise consequences of our actions. According to the more universally held Darwinian model, unconscious, rather than planned, selection is always at work in even the most carefully considered of human projects.

John Langrish has proposed a neo-Darwinian model of evolution that does, I think, permit Dewey and Darwin to inhabit the same page. His position combines Darwin's theory of natural selection with genetics, which appeared after Darwin, as a way of explaining what is carried from one generation to another. Langrish includes in the information carried between generations not just the genetic code of DNA but ideas. As we discussed in chapter 1, this proposal relies on Richard Dawkin's proposal that "memes," or patterns of reasoning, are transferred between generations and thus replicated through time and space. Larngrish thus argues, "[T]he idea that Darwinian change is just 'chance' is wrong." Rather, "[human] striving has to be seen as a necessary but insufficient factor in Darwinian change" (Dawkins 1976, 11). This is to say that striving to make things better is an essential human characteristic, but it does not ensure that change will happen in the direction we intend. Progress is not assured by striving, yet striving does influence not only history but also the environment to which human biology adapts (Langrish 2004). A neo-Darwinian position would, then, accept the theory not only of *genetic* transfer but *memetic* transfer as well (Dawkins 1976; Wikipedia 2005).

To argue that humans and nature coevolve is no longer a controversial proposition except among the religious faithful, but to argue that technologies evolve along with humans and their environment is still a matter of debate. Technological determinists argue that it is technology that leads humans and nature down a particular path of development. In contrast, technological voluntarists argue the opposite: that society decides which technology will take us where we want to go. Both technological determinists and voluntarists, however, are prone to argue that technologies are artifacts, not biological entities, and thus not subject to the Darwinian laws of natural selection. If we accept, however, the notion that technologies are material "memes," or reified ideas that both reflect and determine human striving, the idea that technologies coevolve along with humans and the environment is less problematic. This is the view supported by this investigation (Basalla 1988; Brey forthcoming).

The coevolution hypothesis does not assure that the technological codes for "skyscrapers" developed in Frankfurt, for example, will be repeated and developed along the story lines we now associate with sustainable development. Other stories told by other citizen groups will certainly appear and deflect the trajectory of history desired by the advocates of sustainability. This much has already been documented in chapter 4, but historical contingency in a neo-Darwinian context does not discount the agency of human striving and reasoning. Far from it. Citizens must, however, recognize both the advantages and fallibility of human rationality and thus subject their humble projects to constant experimentation, evaluation, and development. The real successes of Austin, Curitiba, and Frankfurt demonstrate that rapid cycles of action and reflection tend to speed up the coevolution of society, nature, and technology.

6.6 TECHNOLOGICAL CODES AND HUMAN HABITS

Although we might convince a neighbor that recycling her garbage, for example, is a good thing to do in principle, there is no guarantee that she will actually make a practice of doing so. There is not a perfect correlation between what we think and what we do—the complexities of getting through the day intervene against our best-laid plans. Nor will ordering our neighbor to recycle her garbage by imposing fines or other draconian penalties guarantee changed behavior if the old behavior remains more attractive or if she is rich enough. We might, however, stimulate changed behavior by making it more attractive than our current unsustainable habits (Brand 2005). This logic might suggest that all we need do is summon sustainable behavior by making it irresistible. For some this suggestion will reek of social engineering, a concept very much out of tune with our time, but even if we took a legislative route seriously it is not likely that technological codes that positively reinforce sustainable behavior would immediately come into being.

Elsewhere I have argued that technological codes pertaining to the built environment can be historically classified as three types: *tacit*, *legislative*, and *industrial*. Tacit, or unspoken, codes are those that unconsciously bind citizens to the customary or vernacular practices of their community. This mode of regulating behavior has, of course, lost influence as societies have become more self-conscious and diverse. The second type, legislative codes, appeared in Byzantine Roman and Islamic societies by the sixth century and includes explicit civil laws that guard public resources against private exploitation or carelessness. It is this tradition, first incorporated in the West as the Greatest Happiness Principle of the Utilitarian philosophers, that still governs the built environment under the doctrines of police power. The third type of code, industrial, was developed by twentieth-century governments and industry to standardize modern building materials and processes. Such standardizing codes should be understood as economic in origin and were designed to optimize exchange value across political jurisdictions and are linked to the general process of modernization through which the tacit knowledge of the artisan was supplanted by the formal knowledge of the engineer.

The crafting of modern legislative and industrial codes is not purely a matter of science, efficiency, or economics. Rather, it is a highly contested social process through which societies decide how we will live together and in relation to nature. New codes do not simply show up but are negotiated—usually by the parties who have the greatest economic and political interest in the outcome. This logic suggests that citizens would be deeply engaged in code making because controlling the design of artifacts is, to some degree, to control how we live. Sadly, however, this is not the case. Instead, technological codes are generally constructed and implemented by those very technocrats and manufacturers who will profit by altered standards (Moore 2005).

The case of Frankfurt's Commerzbank provides a refreshing example of how code making might be the product of highly democratic and reflective public talk rather than technocratic mandate as in Curitiba. Even if life-enhancing codes are not easily or quickly constructed, the evidence suggests that such codes lead to new social habits that can be attractive to citizens and effective in achieving sustainability and lead to resource efficiency.

Their appearance in the European Union is one indicator that the social values that drive code making in that region have evolved in a desirable direction.

6.7 TECHNOLOGICAL AND "WICKED" PROBLEMS

The notion that there exist "wicked" problems that defy the logic and methods of scientific or technocratic rationality is generally credited to Horst Rittel. In his analysis, urban societies are filled with intractable social problems such as racism, environmental degradation, and suburban sprawl that are simply confounding, or "wicked." For the purpose of this investigation, sustainable development is a classical example of a wicked problem. Rittel and his colleagues have offered six characteristics of wickedness:

1. You don't understand the problem until you have developed a solution. Indeed, there is no definitive statement of "The Problem." The problem is ill structured, an evolving set of interlocking issues and constraints.
2. Wicked problems have no stopping rule. Because there is no definitive " Problem," there is also no definitive " Solution." The problem-solving process ends when you run out of resources.
3. Solutions to wicked problems are not right or wrong, simply "better," "worse," "good enough," or "not good enough."
4. Every wicked problem is essentially unique and novel. There are so many factors and conditions, all embedded in a dynamic social context, that no two wicked problems are alike, and the solutions to them will always be custom designed and fitted.
5. Every solution to a wicked problem is a "one-shot operation"; every attempt has consequences. As Rittel says, "One cannot build a freeway to see how it works." This is the "Catch 22" about wicked problems: you cannot learn about the problem without trying solutions, but every solution you try is expensive and has lasting unintended consequences that are likely to spawn new wicked problems.
6. Wicked problems have no given alternative solutions. There may be no solutions, or there may be a host of potential solutions that are devised and another host that are never even thought of (Rittel 1973).

The type of reasoning articulated here so nicely is what I referred to as *abductive*. It is neither deductive (beginning from principles) nor inductive (beginning from data), but is situated, like the judge and the designer, within the complex conditions of everyday life where meaning is created by purposeful action and its consequences.

		Goals	
		Agreed	Not Agreed
Technologies	Known	Standard engineering practice	Austin's water-quality wars
	Unknown	Curitiba's incremental projects	Frankfurt's banking district

Figure 6.1. Conventional and "Wicked" Problems

One way to understand wicked problems is that neither the goals of action nor the technologies best suited to resolve them are clear at the onset. This relation between goals and technologies, illustrated in figure 6.1, is helpful in plotting how the cities studied employed different types of reasoning in solving their wicked problems (Blanco 1994).

The point of this matrix is to illustrate examples in each of the three cities studied where nontechnological reasoning led to innovation. The first cell (upper left) establishes the norm for standard engineering practice in which technological rationality proceeds by agreeing on goals and employing known technologies to achieve them. A routine example in any city might be the maintenance and repaving of streets, which can be predicted by traffic data and known wear characteristics of various materials. This style of reasoning proves inadequate, however, when either goals are not agreed on or technologies are not available to achieve them. In such cases, most cities solve problems either by contriving limited agreements or by applying inappropriate technologies.

The second cell (upper right) characterizes Austin's water-quality wars as a case in which principal interest groups could not agree on goals, but a single technology was put forward as the solution. In my reconstruction of Austin's story line, the goal of environmentalists was to conserve water quality and the opposing goal of rugged individualists was to conserve property rights. What is a bit surprising is that within the limits of this opposition, only one technological approach was offered to solve the problem, which was to legislate limits on pervious cover by buildings and paving in the Edwards Aquifer recharge zone. Although varying limits were in fact zoned into existence for each of Austin's forty-six watersheds—which was at the time considered a significant innovation—the lingering lack of agreement about goals subverted the effectiveness of the legislation discussed in chapter 2.

The third cell (lower left) characterizes Curitiba's incremental projects as a case in which goals were agreed on but no technologies were available to achieve them. A majority of citizens, planners, and the military government easily agreed that the city needed an infrastructure that would maintain what I called earlier the "civic economy." The problem was that the known technologies that could provide such services were far more expensive than any Brazilian city could hope to afford. Innovation in this case was achieved by adopting unknown technologies developed by the city's creative architects. Here I must expand an argument made by Schwartz that I have already cited in chapter 3—that architectural- or design-thinking is surprisingly more effective when confronted with the indeterminacy of wicked problems than is the technological rationality typically employed by engineers and economists (Schwartz 2004). This logic suggests that design reasoning, as distinct from analytic rationality, is a form of abductive reasoning because it takes place within a social, material, and ecological context.

The fourth cell (lower right) characterizes Frankfurt's redevelopment of the banking district as a case in which there was initial agreement on neither goals nor technologies to satisfy the conflicting demands of activists or bankers. In this case, social learning took place not so much in angry debates as through a prolonged process of testing various design solutions. When the red/green coalition government finally proposed redeveloping the banking area by employing new environmentally friendly technologies and previously unknown mixed-use building types, both sides discovered that they could agree on the specific actions of the redevelopment program even if they saw different principles as motivating the redevelopment effort. The point here is that, as Rittel proposed in 1973, "you don't understand the problem until you have developed a solution." The citizens, bankers, and planners of Frankfurt finally managed to resolve, mostly to their mutual satisfaction, one wicked problem only because they were willing to engage in public talk and various design experiments until they saw an unexpected solution come into view. Foster's design for the Commerzbank became a solution by incorporating new values as attractive ways of living only imagined abstractly by citizens.

Larry Hickman, following Dewey, argues, "A culture which permits science to destroy traditional values but which distrusts its power to create new ones is a culture that is destroying itself" (cited in Hickman 2001, 65). Turned inside out, Hickman's sobering observation would hold that a culture that renders traditional values undesirable by offering more attractive alternatives is a culture that is renewing itself.

6.8 QUALITIES AND CONSEQUENCES OF ACTION

Architects and planners have long debated the relative merits of two "design attitudes," which philosophers refer to more formally as *deontic* and *consequential*. The deontic attitude grants moral preference to particular qualities of human action, such as bravery, simplicity, or generosity. In similar fashion, designers have historically granted preference to such qualities as "classical," "functional," or "regional." The consequential attitude, by contrast, grants moral preference to actions that result in the greatest social good (Ventre 1990). Both attitudes turned up in the study of our three cities.

In Austin, Bill Bunch, director of the Save Our Springs Alliance, adopted a deontic attitude in arguing that "natural" conditions are inherently better than artificial or manmade conditions. On these grounds he holds that development of previously undisturbed land in the fragile Hill Country ecosystem is morally unjustifiable because nature would be compromised—it would become "nonnature" or a human artifact. Even if we could demonstrate that large cities could be designed to be more ecologically robust and life-enhancing than untouched nature itself, Bunch would still argue that the original version holds innate value and on that basis should be left undisturbed by humans. Considered in isolation from the messiness of the world it is hard to disagree with the deontic position—natural things do have integrity. On the basis of deonticism we would, then, try to convince our fellow citizens that the route to the sustainable city is to suppress population growth and make nonnature (that is, cities) as dense as possible.

In contrast to Bunch, Frankfurt city councilor Danny Cohn-Bendit adopted a consequentialist attitude. He initially argued, for example, that no building in the city should exceed five stories in height. The implicit value in his argument was that a five-story limit was consistent with both the historic land-use patterns of the city and with human scale. To violate those norms was, in his initial assessment, to violate the tacit agreements that make Frankfurters, well, Frankfurters. Following more than ten years of debate and design experimentation, however, he publicly advocated for a building of ten times that height—a remarkable fifty stories! Where some Frankfurters interpreted this change of position to be "unprincipled" and therefore unethical, others saw in Cohn-Bendit's shift a changed understanding of Frankfurt's political, economic, and ecological situation. Because he came to see his city as one always in the making, rather than one bound by the tacit agreements of previous generations, he could understand historic land use patterns and human scale as being developmental conditions rather than fixed principles.

These contrasting examples suggest that getting interested citizens in any locale to agree on what is right, good, or true before taking action is a project with dismal prospects. It is highly unlikely that those who hold deontic values, like Bunch, and those who hold consequentialist values, like Cohn-Bendit, could ever agree on their metaphysical assumptions. In the face of such a principled standoff, pragmatists generally agree that success in solving problems is far more likely if we give up the project of asking people to *think* alike and consider what it is that we might *do* together in pursuit of individual goals. In purely practical terms, this means deciding which technologies we are going to use to solve problems to which we can give mutual priority.

This consequentialist logic suggests that individuals must be willing to set aside their incompatible metaphysical beliefs as private disagreements so that we might collaborate in actually *doing* something about current conditions other than debate the meaning of words. Implicit in this proposal is that what we *think* is actually less important than what we *do*. Although this proposal tends to outrage many because of its seeming disregard for the seriousness of ideas, it delights others because of its inclusivity. Nevertheless, by letting people determine the meaning of sustainability for themselves, rather than insisting on splitting hairs or arguing how many angels can stand sustainably on the head of a pin, more people will be drawn into action and thus more creative possibilities will be made available. In the end, it does matter how we define sustainable urban development, but we should also recognize that the meaning of sustainability will become clearer, and mature in the process of aspiring toward it (Thompson 2004). This way of thinking is not relativism but that of pragmatist epistemology.

This logic takes us back to what Jonas Rabinowitz, Curitiba's former information director, referred to as that city's "action scripts." In chapter 3, I distinguished Curitiba's action scripts from what many outside observers mistakenly understood to be the city's success in "comprehensive planning." The difference is that Curitiba's action scripts are focused short-term projects of limited scope, and comprehensive planning is, in contrast, general and long-term. Comprehensive planning is typically driven by the articulation of (deontic) principles that are fixed before the process of implementation begins.[4] The *AustinPlan* of the 1980s is a good, though failed, attempt at comprehensive planning. Rather than being understood as a "project" or "campaign," the *AustinPlan* was understood by its supporters as a political movement—a theoretically inspired and comprehensive change of social organization. Austin's planners and citizens laboriously worked out a detailed master plan for the city as a whole and then, only after all the details were agreed on, did they consider how the plan might be implemented. My assessment is that the *AustinPlan* failed for two reasons: first, because conceptualization was abstract or idealized; and second, because the method of implementation was never even discussed until the plan was complete. In incremental planning, as in Curitiba, there were no grand plans, only humble campaigns of short duration. This manner of planning is continuous rather than episodic (as in Austin) and constantly adjusts the goals of the next campaign on the basis of lessons learned or consequences of the last one.

In summation, I will argue that, on the basis of studying our three cities, planning processes based on deontological principles are generally less successful than consequentialist ones, but this argument cannot be generalized lest it too becomes a fixed and thus inflexible rule.

6.9 TRUTH AND EVENTS

> Politics is not the application of Truth to the problem of human relations but the application of human relations to the problem of truth. Justice, then, appears as an approximation of principle in a world of action where absolute principles are irrelevant.
>
> —Benjamin Barber, *Strong Democracy*

If absolute principles are, as Barber argues, irrelevant in the conditions of everyday life, spilling ink to consider the existence of "truth" must surely be a waste of resources (Barber 1984). My rationale in this section is, however, not to decide such a primary philosophical question but to articulate an attitude toward interpretation that derives from the investigation of our three cities. An attitude, as I use the term here, is a kind of tool—it helps to get work done.

Interpretation through pragmatist lenses does not discount the possibility of "truth." It does, however, discount the possibility of "Truth." The distinction that I would like to make here is between humble truth as an event, and heroic Truth as a fixed principle. Modern positivists like Marx claimed that science can reveal the objective structures of history—race, gender, and class for example—as the "Truth." Postmoderns such as Derrida, however, contend that all truth claims are contingent because they are socially constructed and therefore relative. For orthodox moderns truth is absolute and for postmoderns it does not exist at all. Following William James, I find a third position more helpful if not more certain—that "Truth" is best understood as an *event*.

For James, "Truth *happens to an idea*. It *becomes* true, is *made* true by events. Its verity *is* in fact an event, a process: the process namely of verifying itself" (James 1907, original emphasis). The idea that truth is evolutionary, rather than fixed, was also adopted by John Dewey. In Dewey's assessment, it is humankind that pushes development of the world, not some kind of historical teleology or divine plan but humans coming to consciousness through social discourse. It is social discourse, then, that serves to stimulate human self-reflection on which evolutionary history depends (Hickman 2001).

In each of the stories reconstructed previously some respondents adopted what I will refer to as a developmental attitude toward cities. In Austin, Bob Paterson argued that central Texans would have to develop new values in order to survive. In Curitiba, Lerner equated cultural evolution with hope, and in Frankfurt it was Helmut Bosch who most clearly articulated a developmental attitude toward reality. My point is not to argue that these particular individuals were extraordinary leaders in their respective cities, but to argue the reverse, that their attitudes reflect what was happening in their respective cities generally—at least for a time.

Those who argue that human nature is fixed also argue that stories in themselves do not drive history. This is to say that individuals in isolation cannot fabricate possible futures that will have much effect on history. I agree. But there is a difference between individual hopes and socially constructed horizons of possibility. These are what Dewey called "ends-in-view," a moment when society lends its collective imagination to ending an economic depression or putting a man on the moon. It is these collective plans that become the means of reconstructing both social consciousness and the material city. There is, then, a continuum of "ends-means" in collective storytelling that is very unlike individual hopes or fantasies.

Andrew Feenberg's notion of "civilizational change," or the adoption of new paradigmatic values by society, will help to make the difference clear. His argument is that such changed cultural values show up in the world as new technological codes (Feenberg 2002). The appearance of building entry ramps for handicapped citizens, for example, is an indicator that society no longer accepts the exclusion of these citizens from public life. The narrative that enabled such new technological codes to be legislated evolved over at least two decades and became the "ends-in-view" shared by a majority of citizens. In this case, the physical reconstruction of the urban lifeworld evolved significantly in a rather short period of time because a powerful new story line was constructed through social discourse and subsequent reflection. Dewey's argument is that such reconstructions of the lifeworld, based on changing values, add up over time to evolutionary changes in humans, and so on.

This example suggests that sustainable urban development might show up and endure as the dominant story line of a particular place—it might become "true." The reason why Austin, Curitiba, and Frankfurt are so important is because these cities demonstrate how public talk and self-reflection have become a life-enhancing force in evolutionary history.

6.10 MODELS AND LISTS

In chapter 1, I argued that conceptual models of sustainable development (like the "planner's triangle" implicit in the Brundtland Report) and lists of "best practices" (like LEED, Leadership in Energy and Environmental Design) were of heuristic value to those wishing to better understand public talk about sustainability but that, in the end, they were distractions from the pursuit of developing and refining local story lines. If you understand sustainable urban development as an ideal model, I argued, you will encounter obstacles only when applying the abstract model to a real city. In contrast, if you understand sustainable development as a process of local storytelling you will encounter opportunities that relate local conditions to global structures. It is not surprising that the reconstruction of the three cases studied in this book seems to have reinforced that judgment. The thick stories that respondents told about their cities seem resistant to any common conceptual structure—save the analytical method I employed to document *political, environmental,* and *technological talk* in each city. At this point in the analysis, it would only be circular logic to argue that the analytic method employed proved itself to be true, so I will resist that temptation.

I do, however, wish to reinforce another argument made briefly in chapter 1 because it raises a methodological problem that I should address before concluding. The argument made at the outset is that conceptual models are examples of deductive logic and that lists of best practices are examples of inductive logic. On the surface, this is hardly a controversial or problematic claim. Models rely on a priori reasoning and lists of best practices aggregate practices that have historically proven helpful. The argument becomes controversial only by arguing, as I do earlier, that these traditional forms of reasoning are less well suited to solving wicked urban problems than is abductive reasoning employed in the construction of urban story lines. To be clear, my argument is not that deduction and induction are unhelpful but rather that abduction can be more helpful in producing hypotheses that will solve wicked problems, with the assistance of deduction and induction.

The difficulty with this claim is found in the cases selected for study. By selecting three cities on three different continents with such dramatic ecological and cultural differences, the analysis tends to obscure the similarities that surely exist between cities within a particular region. In other words, my analysis has been *inter*regional not *intra*regional. I should then consider whether it might be reasonable to employ a model or list, for instance, within Germany, southern Brazil, or the American Sun Belt, where intraregional differences might be less dramatic.

The most direct answer I can offer to this question is that "it depends." If we return to the brief histories of each city reconstructed earlier, it is very clear that Austin and Curitiba are both anomalies within their regions. Even if the ecologies of Austin and San Antonio, or Curitiba and Porto Alegre, for example, were similar, their political cultures could not be more different—and it is the political cultures of cities, more than their ecologies, that frame modes of implementation. Frankfurt may be slightly less anomalous in its region, yet significant cultural differences remain between it and, for example, Düsseldorf. As I argued previously, local conditions are not an obstacle to be overcome by universal principles as much as an opportunity to hang universal ideas on stories and practices that are already part of people's everyday lives. At the scale of recycling garbage, there may be little difference in how such a universal idea is implemented within a region, but at the scale of architecture and urbanism, *how* citizens build in Frankfurt and in Düsseldorf might be very different indeed. So, in the choice between acting on models of sustainable development, lists of best practices, or local story lines, it depends on the local situation. On the basis of this study, however, helpful models and lists appeared only in cities that aspire to sustainability *after* powerful stories had been locally constructed. This is to say that models and lists are at best analytic and not causal.

6.11 MOBS, CLIENTS, AND CITIZENS

I argued in chapter 1 that all three of the cities studied have developed different kinds of democracy by engaging in political, environmental, technological, and other kinds of public talk. None of the cities are pure examples of any one particular kind of democracy, but it is fair to say that Austin leans toward populist democracy, Curitiba toward technocratic democracy, and Frankfurt toward strong democracy. I will briefly summarize each in turn.

Students of political culture argue that political populism tends to gravitate toward one of two poles: toward tribal forms of mob mobilization or toward static compromise between competing interest groups (Barber 1984; Fischer 2003). The successive Vargas regimes in Brazil (1930–1954) are an example of populist mob mobilization and Austin's regime of sustainability was an example of populist static compromise. Neither direction, I will maintain, lends itself to long-term sustainable urban development because both feed on irrational forms of public talk in which public learning cannot take place. This claim is made on our understanding that participation in a mob, or even in a quiet neighborhood cleanup campaign, does not qualify as "public talk" as I have employed the term herein. Rather, public talk requires rational deliberation, and it is only through such deliberation that social learning takes place. Neither irrational mobs nor nondiscursive action contribute to public talk. Participation in itself is a necessary, but insufficient, condition to catalyze democracy.

Curitiba's regime of sustainability was technocratic in nearly every sense of the term. Self-appointed elites managed to keep their clients happy by managing the municipal infrastructure efficiently. In light of the substantial evidence, I must agree that technocratic rule from the top can solve real problems for clients. This does not mean, however, that such development patterns can be indefinitely sustained because public learning in this kind of democracy, like tribal collective consciousness, is wanting. In the absence of public learning, clients will not be able to adapt the systems that keep the trains running on time when the managers disappear.

If neither "mobs" nor "clients" are likely to achieve long-term sustainable development, the preceding analysis suggests that "citizens" like those of Frankfurt are our best bet. This claim, however, is open to a substantive critique by those who maintain that there is "surprisingly little evidence" that citizen participation "contribute[s] to the fabric of the social and political environment of sustainability" (Portney 2003, 249). In chapter 5, I did refer to empirical evidence demonstrating a positive correlation between democracy and sustainable development and I suspect that research in this area will continue. In spite of such emergent empirical evidence, however, I will have to agree that support for the hypothesis is not yet overwhelming. Nonetheless, at this stage in the investigation it is reasonable to argue that strong democracy may not cause sustainable urban development, but there is clearly a correlation between the two phenomena (Portney 2003).

This claim relies on two observations—one substantive and one methodological. First, the substantive observation is that the case studies analyzed here do provide qualitative and quantitative evidence that correlates disciplined public talk with conditions that most people associate with sustainability. This is not to say that all public talk in these three cities was disciplined or that no talk was sloppy or suppressed. But it is to say that the story lines constructed in all three cities by citizens included their own participation as a dominant or countervalue. Even in Curitaba, where strong citizenship is far less developed than are implicit contracts between technocrats and their clients, there is growing awareness and talk about citizen engagement—it is valued.

Second is a methodological argument to support citizen participation as an element of sustainable development. Advocates of naturalistic inquiry argue that the kind of local knowledge that leads to the sustainable development of a particular place is "inaccessible to [the] more abstract empirical methods" accepted by positivistic social scientists as evidence (Fischer 2000, 2). "Insofar as many social problems have their origin in a local context—environmental problems being a prime example—knowledge of the local citizen's understanding of the problem is essential to effectively identifying and defining the problem" (Fischer 2000, 217). My point here is that the knowledge essential to the pursuit of sustainability can be invisible to the empirical methods generally used to quantify sustainability indicators but has been highly visible in the ethnographic stories of our three cities reconstructed earlier.

6.12 CONCEPTUALIZATION AND IMPLEMENTATION

In chapter 3, I argued that the Lerner regime of sustainability had developed a "theory of implementation" that was largely responsible for Curitiba's success (Schwartz 2004). If we mean by "theory" a set of flexible principles devised to explain accepted "facts" that can be used to make predictions, then yes, it does seem that Lerner and his colleagues had a theory of implementation. So did the citizens and planners of Frankfurt, if not Austin. In both cases, however, their theories were not formal scientific ones, but were of a particularly informal or ad hoc sort—what we might better refer to as an *operating* hypothesis. Note that the emphasis of that term is put on the operation, or action toward a common goal, not on the precise nature of the relationship between the variables.

If we contrast this attitude to that of ideological purists (Davison 2001), the issue at hand becomes clearer. Where ideological purists require that we all submit to a common theory before we act (lest we act wrongly), the ad hoc planners of Curitiba required only a short-term commitment to a project (lest we not act on a fleeting opportunity). Lerner himself put it succinctly by arguing that "it's necessary to start without all the answers" because the answers will change as the opportunity unfolds. As a calculated way to shock veteran economists who insist on complex long-range planning, Lerner and his colleagues are fond of characterizing their approach to planning as "ready, fire, aim" (Lerner 2006). I should, though, take care again not to paint Lerner and his colleagues too thoroughly as pragmatists. Curitiba's planners and the pragmatists do not share a faith in the value of citizen participation. What they do share is a deep skepticism of "theoretical hallucinations"—the notion that we can think our way through to consequences before action takes place (Rorty 1998).

In the pragmatist view, theory and practice are complementary and integrated activities that constitute "a conversation that constantly adjusts means to ends-in-view, and ends-in-view to means at hand"

(Hickman 2001, 180). This logic suggests that you can never conceptualize how the world should be separate from implementation. This is to say that modes of implementation are theories of conceptualization in disguise.

In sum, the situated attitudes that I have recommended throughout this analysis of the preceding twelve dilemmas can be characterized as developmental rather than absolute or fixed. Solutions to wicked problems—if they are to be found at all—suggest themselves as they come into view. Sustainable development, then, might be understood as unending cycles of action and reflection that move us forward.

Arguing that the insights gained from this investigation might move us forward must sound rather optimistic—optimistic enough to constitute an endorsement for the much contested notion of "progress." This final observation requires that I consider the history of such optimism, and objections to it, before offering twelve corresponding abductive tools as a conclusion.

CHAPTER REFERENCES

AtKisson, A. (1999). *Believing Cassandra: An optimist's look at a pessimist's world.* White River Junction, VT: Chelsea Green.

Bahro, R. (1994). *Avoiding Social & Ecological Disaster: The politics of world transformation.* Bath, UK: Gateway Books.

Barber, B. (1984). *Strong Democracy: Participatory politics for a new age.* Berkeley: University of California Press.

Basalla, G. (1988). *The evolution of technology.* Cambridge: Cambridge University Press.

Benton, T., Ed. (1996). *The greening of Marxism.* New York: Guilford Press.

Bernstein, R. J. (1992). "Heidegger's Silence: Ethos and technology." Pp. 79–141 in *The New Constellation: The ethical-political horizon of modernity/postmodernity.* Cambridge, MA: MIT Press.

Blanco, H. (1994). *How to think about social problems: American pragmatism and the idea of planning.* Westport, CT: Greenwood Press.

Bramwell, A. (1985). *Ecology in the twentieth century: A history.* New Haven, CT: Yale University Press.

Brand, R. (2005). *Synchronizing science and technology with human behavior.* London: Earthscan.

Brey, P. (Forthcoming). "Technological Design as an Evolutionary Process." In *Designing in Engineering and Architecture: Towards an Integrated Philosophical Understanding.* P. Kroes, A. Light, S. A. Moore, and P. E. Vermaas, eds. Berlin: Springer.

Chadwick, E. (1965). *Sanitary conditions of the labouring population of Great Britain.* Edinburgh: Edinburgh University.

Davison, A. (2001). *Technology and the contested meaning of sustainability*. Albany: State University of New York Press.
Dawkins, R. (1976). *The selfish gene*. New York: Oxford University Press.
Dewey, J. (1991 [1938]). *Logic: The theory of inquiry*. Vol. 12. Carbondale: Southern Illinois University Press.
Dryzek, J. S. (1997). *The politics of the earth: environmental discourses*. Oxford: Oxford University Press.
Ehrlich, P. R. (1968). *The population bomb*. New York: Ballantine Books.
Feenberg, A. (2002). *Transforming technology: A critical theory revisited*. New York: Oxford University Press.
Fischer, F. (2000). *Citizens, experts, and the environment: The politics of local knowledge*. Durham, NC: Duke University Press.
Foucault, M. (1977). *Discipline and punish: The birth of the prison*. New York: Pantheon.
Groat, L. and D. Wang. (2002). *Architectural research methods*. New York: Wiley.
Guy, S. and E. Shove. (2000). *A sociology of energy, buildings, and the environment: Constructing knowledge, designing practice*. London: Routledge.
Haraway, D. (1995). "Situated knowledge: The science question in feminism and the privilege of partial perspective." Pp. 175–94 in A. Hannay and A. Feenberg, eds., *Technology & the politics of knowledge*. Bloomington: Indiana University Press.
Hardin, G. (1968). "The tragedy of the commons." *Science* (162): 1243–48.
Harvey, D. (2000). *Spaces of hope*. Berkeley: University of California Press.
Hickman, L. A. (2001). *Philosophical tools for a technological culture: Putting pragmatism to work*. Bloomington: Indiana University Press.
James, W. (1907). *Pragmatism, a new name for some old ways of thinking: Popular lectures on philosophy*. New York: Longmans, Green, and Co.
Langrish, J. Z. (2004). "Darwinian design: The memetic evolution of design ideas" *Design Issues* 20 (4) (Autumn): 2–19.
Lerner, J. (2006). *Tropical green*. Telecast from University of South Florida, Miami, 9 February.
Malthus, T. (1798). *An assay on the principle of population, as it affects the future improvement of society with remarks on the speculations of Mr. Godwin, M. Condorcet, and other writers*: London: Printed for J. Johnston in St. Paul's Church-yard.
Marx, K. (1908). *Das Kapital*, 4th ed. London: W. Reeves.
Meadows, D. (1995). "The city of first priorities." *Whole Earth Review* (Spring): 58–59.
Moore, S. A. (2005). "Building codes." Pp. 262–66 in C. Mitchum, ed., *Encyclopedia of science, technology, and ethics*. New York: Macmillan.
Perleman, M. (1996). "Marx and resource scarcity." In T. Benton, ed., *The greening of Marxism*. New York: Guilford Press.
Portney, K. E. (2003). *Taking sustainable cities seriously: Economic development, the environment, and quality of life in American cities*. Cambridge, MA: MIT Press.
Prugh, T., R. Cosanza, and H. Daly. (2000). *The local politics of global sustainability*. Washington, D.C.: Island Press.
Rees, W. (2004a). "Globalization and Sustainability: conflict or convergence." Working paper, University of Texas Center for Sustainable Development. <web.austin.utexas.edu/csd/index.html>. (accessed 7 June 2004).

Rees, W. (2004b). Personal correspondence, 12 March.
Rittel, H. and M. Webber. (1973). "Dilemmas in a general theory of planning." *Policy Sciences* (4): 155–69.
Rohracher, H. (1999). "Sustainable construction of buildings: A socio-technical perspective." Paper presented at the Proceedings of the International Summer Academy on Technology Studies: Technology Studies and Sustainability, Inter-University Research Center for Technology, Work and Culture. Graz, Austria.
Rorty, R. (1998). *Achieving our country*. Cambridge, MA: Harvard University Press.
Schwartz, H. (2004). *Urban renewal, municipal revitalization: The case of Curitiba, Brazil*. Alexandria, VA: Hugh Schwartz, Ph.D.
Thompson, P. (2004). "What sustainability is and isn't." Working paper, University of Texas Center for Sustainable Development. <web.austin.utexas.edu/csd/index.html>. (accessed 1 September 2004).
Ventre, F. (1990). "Regulation: a realization of social ethics." *VIA* (10): 51–62.
Wikipedia. (2005). "Memetics." <en.wikipedia.org/wiki/Wikipedia>. (accessed 10 December 2005).

NOTES

1. Another way to argue this point is to say that case studies may be generalized "theoretically" but not "literally" (Groat 2002).

2. I am indebted to an anonymous referee of the manuscript for this argument. The ethical problem in this logic emerges from the inability of some scientists to distinguish between what is scientifically possible from that which is culturally desirable.

3. Bernstein and Fischer, also following Aristotle, refer to a similar manner of reasoning as "*phroenesis*," by which they mean the responsibility of every citizen to critically examine the work of their peers for the benefit of society as a whole (Bernstein 1992; Fischer 2000). Donna Haraway has made a similar argument in favor of what she calls "situated knowledge." Her critique of "objective" technological rationality is that:

> All Western cultural narratives about objectivity are allegories of the ideologies governing the relations of what we call mind and body, distance and responsibility. Feminist objectivity is about limited location and situated knowledge, not about transcendence and splitting of subject and object. It allows us to become answerable for what we learn how to see. (Haraway 1995, 185)

Although "abduction," "critical thinking," "situated knowledge," and "phroenesis" should not be conflated as meaning the same mental process called by different names, they are strongly related and collectively document that many scholars have characterized modes of rational thinking that is distinct from classical deduction and induction.

4. Richard Rorty has best distinguished between a "campaign" and a "movement" in saying, "[B]y 'campaign,' I mean something finite, something that can

be recognized to have succeeded or to have, so far, failed. Movements, by contrast, neither succeed nor fail. They are too big and too amorphous to do anything that simple. They share in what Kierkegaard called 'the passion of the infinite'" (Rorty 1998).

SEVEN

Alternative Routes to the Sustainable City

The study of Austin, Curitiba, and Frankfurt confirms that there are indeed alternative routes to the sustainable city. No single path of action prescribed by abstract models or lists of best practices guarantees success in the ellusive pursuit of sustainable urban development. If anything, the cases studied suggest just the opposite: sustainability tends to show up in cities that become self-conscious and practiced in constructing, merging, and reconstructing their own story lines. If there is a general rule, or hypothesis, that can be gleaned from the analysis, it is that sustainability is not a fixed condition but a dynamic meta-story line that links local political, environmental, and technological stories to an attractive plot. It is the plot shared by a majority of citizens that attracts sustainable human practices toward an unfolding horizon of sustainability.

Any positive proposal for a meta-story or meta-narrative will, however, be met with grave skepticism by those postmoderns such as Foucault (1926–1984) or those critical theorists such as Max Horkheimer (1895–1973) who would today probably fear sustainability as only another in the long tradition of totalizing "master-narratives" that have proposed to diagnose the present and prescribe the future in precisely the same manner as did the Enlightenment (Mol 2003). Although postmodern skepticism has much to recommend it, I argue later that it often ends up being a thinly veiled conservative excuse for inaction. My argument is not made to dismiss postmodern skepticism of Enlightenment forms of rationality but to put this way of thinking in a historical context before offering the twelve abductive tools that conclude this study.

7.1 MASTER- OR META-STORY LINES

Having choked on the smoke of industrialization, romantics in Europe and North America were the first to abandon faith in the benefits of human rationality. It was the horror of two twentieth-century world wars, however, that focused postmoderns and critical theorists on the demon within us. From both perspectives, human rationality had become associated with the obscene order and certainty of the future imagined first by British Utilitarians and second by German National Socialists. Foucault's critique of the Utilitarians centered on the modern formation of a "disciplinary society" orchestrated by the illegitimate exercise of power by the state through the organization of economic, "juridico-political," and scientific institutions. In his view, the British public health movement initiated by Bentham, Mill, and Chadwick, which I discussed briefly in chapter 5, exemplified the will of modern technoscience to "engineer the soul" (Foucault 1977). Although Foucault's style of anthropology has received criticism itself in recent years, it still stands as a viable caution against the modern will to bureaucratize the lifeworld first foreseen by Max Weber (1958). In similar fashion, Horkheimer and Theodor Adorno portrayed modern science, particularly the variety employed by German National Socialists, as the will to predict and control nature in a way that would surely end in a kind of universal darkness not unlike that experienced at Auschwitz (Horkheimer 1947).

In the shadow of the postwar period, uncritical dependence on the rational method caused American pragmatists, and Dewey in particular, to be criticized as naïve advocates of action, unreflective about the historical consequence of embracing Enlightenment rationality too closely (Hickman 2001). That critique continues as a response not only to pragmatism as a tradition of thought but also to the proposal for sustainable development studied here. A particularly clear example of that critique is offered by architect and theorist Paul Shepheard, who argues that

> [s]ustainability is usually configured as a piece of critical theory against the American way of life. I am suggesting the opposite: that it is part of the American hegemony's desire for perpetuity, a device for making America last forever. (Shepheard 2003)

In *Artificial Love*, Shepheard suggests that sustainability is the thinly veiled doctrine of overzealous do-gooders intent on making the world safe for democracy in the manner of U.S. president George W. Bush. His characterization of sustainability seems to rely on two related assumptions: first, that sustainable development is a unified discourse with foundational beliefs, and second, that it "is a dream of perfected automation" (2003, 194). With regard to the first assumption, the analysis of the various and often competing discourses associated with sustainability in chapter 1 and, as documented in the case studies earlier, seems adequate to undermine Shepheard's assumption. To be clear, however, I do agree that there are some advocates of some sustainability discourses who do fit

Shepheard's characterization of militant "fundamentalists." Nevertheless, the existence of some such people does not make it reasonable to essentialize all sustainable advocates as being of a single mind (Guy 2001; Guy and Moore 2004). As Horkheimer himself argued in the 1940s, the essentialization of people and ideas is usually the project of a demagogue.

Second, the characterization of sustainable development as being a "dream of perfected automation" is an interesting one. As I argued in chapter 4, the Frankfurt bankers who push "ecological modernization" may actually understand sustainability in such mechanistic terms, but, if anything, Shepheard's metaphor is one well suited to the mechanical paradigm of production that most advocates of sustainability have rejected in favor of other metaphors, as documented in chapters 2 through 4, related to biological systems that are continually redirected by evolving conditions. Here too, however, Shepheard's logic does have some value because it cautions us, as did Foucault and Horkheimer forty years ago, against confusing the determinacy of technological rationalism with the indeterminacy of life-enhancing evolution. It is this caution that contributes to my rejection of models and lists as prescriptive paths of action. Still, as a whole, Shepheard's critique of sustainability suffers from an ideological blind spot that is unsupported by the evidence.

At the heart of the postmodern critique of sustainability and pragmatism is the notion that "progress doesn't improve the world"; it just makes it more complex (Shepheard 2003, 111). Rather, viewed through postmodern lenses it is the well-intended but always faulty desire of self-selected individuals to improve things that has been the source of modern injustice and environmental degradation. Postmoderns have learned, it seems, that modern technologies "bite back" (Tenner 1996). From this vantage point, the very idea of sustainable development appears synonymous with the contested modern idea of progress.

But there is another way to view the situation. Richard Rorty has been highly critical of the Left for having become—like Shepheard, I argue—"spectatorial and retrospective" (Rorty 1998). By this Rorty means the Left, having recoiled from active engagement in politics after the Vietnam War, has ceased being the Left. Instead, it has become a group of well-educated observers who watch the world unfold in shame—who "prefer knowledge to hope" (1998, 37). With regard to Shepheard, as is the case with many architectural theorists, fear of the unintended consequences associated with trying to actually improve the conditions suffered by some of our fellow citizens has led to paralysis and a convenient withdrawal into aesthetic discourse. In his review of Shepheard's book, Stephen Fox observed not only that its logic is "self indulgent" but "no issues are resolved. No ideas are refined. At the end, the stories simply start over again" (Fox 2004). The stories told by postmoderns such as Shepheard are surely entertaining but also deceptive. In painting the advocates of sustainability as radical fundamentalists set on prediction and control of the world they paint themselves as the avante garde, thus concealing their own conservative social values behind the mask of progressive art.

The thrust of my argument is that the very real threat of Enlightenment science, at least the kind embraced by National Socialists, has not been the use of too much rationality as the postmoderns would have it, but too little (Herf 1984). This is to say that not all forms of rationality, as I have distinguished between them earlier, lead to the same dark future. This logic, however, should not be construed to mean that the pragmatists always got it right.

In the assessment of Andrew Feenberg, "Dewey had intimations of the technocratic threat to democracy contained in the extension of technology," but he failed to adequately consider (or express) his reservations about large-scale corporate-dominated industrial production (Feenberg 2003). From a contemporary perspective, we might surmise that had Dewey witnessed the emergence of the sport utility vehicle—the ubiquitous SUV— as America's most popular car, or the domination of the retail sector by Wal-Mart, he might have been less enthusiastic about the benefits of rational planning in the market. Such Monday morning quarterbacking, however, does not permit us to overlook Dewey's lack of caution toward the perceived benefits of technological rationality. It will suffice to say that the caution provided by Horkheimer, Foucault, and others is well grounded in the historical context of the postwar period. All the same, that caution should not require us to embrace mysticism on the rebound, give up politics, or be prevented from employing the tools of abductive reasoning that have developed over the past century.

7.2 TWELVE ABDUCTIVE TOOLS

At this late point in the investigation the reader might still reasonably ask: how is a list of "abductive tools" different from a list of "best practices"? They are, after all, both meant to direct action. They are both "memes" in the context discussed in chapter 1—"self-replicating units of cultural evolution" (Wikipedia 2005). In response, I will have to agree that lists are lists. There are, however, critical differences.

I argued previously that, where lists of best practices function to cut off conversation by telling people which technology to use, the abductive tools offered next function to open new conversations designed to question technological choices. The primary difference is that best practices rely on the authority of technocratic sponsors at the top to tell us what is right and abductive tools rely on the local intelligence of participants at the bottom to test new possibilities.

In the culture of city planning, it is common for insiders to characterize their discipline as an "applied science," which suggests that planners merely apply what is already known. This is the assumption behind lists of best practices. From the situated perspective developed earlier, however, one learns only through the endless cycles of experimental action and reflection. The purpose of abductive tools, then, is to liberate planners, architects, engineers, and—most of all—citizens from stale assumptions produced by well-meaning technocrats. This logic suggests that defining sustainable urban development in fixed scientific terms to be applied at a distance at some time in the future is a far less desirable goal than being able to recognize the opportunities and conditions that evoke community action toward living sustainably. This judgment derives from the argument made earlier that the meaning of sustainability will continue to change with ever-evolving ecological and social conditions. In the end, it is less important what we think sustainability might mean than what we are willing to do to achieve it. Of course, neither side of the dichotomy constructed here will, or should, dominate what we think or do. The point is that abductive tools are required to hold list makers of any stripe accountable for the direction and consequences of cultural and environmental change.

Holding the institutions of technoscience accountable can be achieved only by proliferating the means of thinking, not by relegating thinking to yet more experts. Unfortunately the institutions of public learning at all levels in the United States and Brazil, if not Germany, have increasingly moved toward disciplinary specialization and away from the study of "civics," or the system of freedoms and responsibilities that tie the disciplines and members of a society together. In such atomized cultural climates highly entropic stories, cultural ideas, or memes can proliferate precisely because there is little rational deliberation in everyday life to challenge them. One need only turn on the television or visit the local Wal-Mart to experience this phenomenon.

If we accept the notion that cultural ideas are passed along through stories of our own making and telling, we must also accept the proposition that we are each engaged in the project of cultural and environmental evolution. The stories and counter-stories that we tell have some, perhaps infinitesimally small effect on the alternative routes that might be available for future generations to select. Our heirs will inherit not only the DNA and environmental conditions that we leave behind, but also our ideas—our stories. This is also to argue that prospects for the future are conditioned, at least in part, by rational mutation—how we reinterpret the stories and material conditions left to us. The following propositions are offered as provisional means toward that end:

1. Regimes of sustainability will tend to show up in culturally diverse spaces where coalitions of environmentalists and social justice advocates redescribe dominant story lines in ways that are attractive to most citizens.
2. Projects are likely to be considered successful by more people when experts depend on citizens to define them.
3. Efficient design will optimize what is technically possible, but effective design will optimize what is socially desirable.
4. Natural resources do have finite limits, but these can be stretched by human mental labor.
5. Because humans, nature, and technologies coevolve, changes in one of these variables can never be studied in isolation.
6. The appearance of new technological codes reflects changed social values and stimulates changed social habits.
7. "Wicked" problems can be solved by employing experimental design thinking, not by sticking with the same scientific assumptions, traditional values, and social habits that created them.
8. Be concerned with the consequences of actions more than their qualities—how brave, simple, or generous they are.
9. It is not particularly helpful for citizens to be concerned with scientific "Truth," but it is very helpful to figure out what it is that we can *do* together to solve common problems.
10. Conceptual models and lists of best practices are of some heuristic value but tend to divert attention away from local opportunities for action that derive from local story lines already related to sustainable development.
11. Although irrational mobs and disciplined clients can both contribute to sustainable conditions in the short run, rational deliberation among citizens contributes most in the long run.
12. Methods of implementation are theories of conceptualization in disguise.

How these propositions are "operationalized" (as scientists say) or put to work (as ordinary citizens say) will depend on the situations into which these words get thrown. There is no formula that will guarantee a direct route to the sustainable city. Rather, it is the responsibility of citizens to tell attractive stories—to catalyze preferred situations by redescribing existing ones.

CHAPTER REFERENCES

Feenberg, A. (2003). "Pragmatism and critical theory of technology." In *Techne* 7 (Fall): 42–49.

Foucault, M. (1977). *Discipline and punish: the birth of the prison*. New York: Pantheon.

Fox, S. (2004). "Review of *Artificial Love*," *Books and buildings*. Lecture delivered March 4 at The University of Texas at Austin, School of Architecture. Unpublished manuscript, collection of the author.

Guy, S. and G. Farmer. (2001). "Reinterpreting sustainable architecture: The place of technology." In *Journal of Architectural Education* 54 (3): 140–48.

Guy, S. and S. Moore, eds. (2004). *Sustainable architectures: Cultures and natures in Europe and North America*. London: Routledge.

Herf, J. (1984). *Reactionary modernism: Technology, culture, and politics in Weimar and the Third Reich*. Cambridge: Cambridge University Press.

Hickman, L. A. (2001). *Philosophical tools for a technological culture: Putting pragmatism to work*. Bloomington: Indiana University Press.

Horkheimer, M. and T. Adorno. (1947). *Dialectic of Enlightenment*. Amsterdam, Netherlands: Querido.

Mol, A. P. J. (2003). "The environmental transformation of the modern order." Pp. 203–24 in P. B. Thomas, J. Misa, and Andrew Feenberg, eds., *Modernity and technology*. Cambridge, MA: MIT Press.

Rorty, R. (1998). *Achieving our country*. Cambridge, MA: Harvard University Press.

Shepheard, P. (2003). *Artificial love: A story of machines and architecture*. Cambridge, MA: MIT Press.

Tenner, E. (1996). *Why things bite back: Technology and the revenge of unintended consequences*. New York: Knopf.

Weber, M. (1958). *The city, by Max Weber*. Translated and edited by Don Martindale and Gertrud Neuwirth. Glencoe, IL: Free Press.

Wikipedia. (2005). "Memetics." <en.wikipedia.org/wiki/Wikipedia> (accessed 10 December 2005).

APPENDIX

Methodology

In chapter 1, I provided an overview of the methods used to collect and interpret the data gleaned from the case studies. The purpose of this appendix is to place those strategic methods within an overview of the various methodologies, or paradigms of inquiry, that are used by other analysts.

In arguing, as I do previously, that we might be better off by shifting our priorities from thinking to doing, I do not mean that research is no longer necessary or that we should act without reflection. To the contrary, I mean that doing is a component of thinking and thinking of doing. Perhaps oddly, I expect that traditional positivists would agree with this logic, at least so far. Experimentation is, after all, nothing more than the study of what nature does. It is the doing of nature that is the subject of so much scientific thinking. Disagreement occurs, however, when positivists assert, as they have traditionally done, that they themselves can be isolated from the doings of nature under study by controlled experimental methods.

The problem with applying purely experimental methods to the study of enormously complex social phenomena like cities, as traditional Marxists attempted to do, is that meaning changes over time. Humans are thrown into a world in which the social meanings associated with landscapes, cities, and buildings seem more or less stable—at least at first. Individuals and societies develop quickly, however, as do their interpretations of artifacts and conditions. For example, we saw in chapter 4 how the citizens of Frankfurt self-consciously changed the social meaning of "skyscrapers" in a mere twenty years. The problem for experimental research in this situation is that there is no anchor for an objective interpretation of reality (Fischer 2003).

Social science has tried to subvert this problem by introducing case studies and ethnographic methods as hypothesis builders, as I have done in this book. The operating hypothesis, built upon phenomenological interpretation, can then be further tested by traditional empirical means. The narratives that experimental research provides us are, however, always about the past. They document how a limited number of variables interact under controlled conditions. It is true that we are free to project antecedent stories into the future, but we do so at our own risk because conditions evolve. A crucial difference between traditional experimental methods and those proposed by Peirce is that the former is focused upon antecedent phenomenon and the latter upon consequent phenomenon. The difference is not a minor matter of interpretive method, but a major matter of epistemology. In the case of positivistic empiricism, we investigate the past in the hope that our generalizable knowledge will be applicable in all places for some period into the future. In the case of pragmatism, analysts investigate future possibilities for action in a particular place. In the end, these are very different but not entirely incompatible goals and methods.

Traditional empirical science depends upon a "correspondence theory" of knowledge in which scientists have the ability to employ language that represents nature exactly as it is. In this worldview, reality exists and can be apprehended by objective observers. Some contemporary social scientists retain this belief, but many others have abandoned it in favor of a "coherence theory" of knowledge in which the investigator expects only that the language employed by any particular social group is internally consistent. In this worldview, reality is socially constructed and can be understood not objectively as in traditional science but in a transactional, or political, relationship between the observer and the observed. Table A.1 includes a very general overview of five epistemological positions that are common in contemporary research. Of these, only traditional positivists hold to a pure correspondence theory of knowledge. Other analysts may include other categories or describe their assumptions differently, so these categories should be understood as a heuristic, rather than a comprehensive list. My purpose in including it here is to place what I understand to be a pragmatic approach to research within a historical spectrum.

Postpositivists, as placed in table A.1, can be characterized as humble positivists. Where positivists argue that objective scientists can apprehend reality through the symbolic methods and languages of science, postpositivists understand that there is never a perfect correspondence between language and nature. Postpositivists do argue, however, that the gap between real conditions and representation of them can be reduced to insignificance by objective methods.

Both positivists and postpositivists rely upon a classical Cartesian worldview in which scientists (subjects) observe the world (objects) at a distance. The assumption of this dualistic picture is that science is "prepolitical," or objective. Donna Haraway's critique is, however, that by seeking certainty via the methods of deduction and induction, positivists miss the embedded social nature of knowledge altogether (Haraway 1995). Simply put, she rejects the self-serving claim of positivists that the

Table A.1 Traditions of Knowing

Tradition	*Position toward Knowing*	*Appropriate Research Methods*	*Exemplars*
Positivism	Reality exists— language can correspond directly to nature	Quantitative/ empirical science	Comte, Marx
Postpositivism	Reality exists— language can effectively approximate nature	Quantitative/ empirical science Quantitative/	Popper (sometimes Dewey)
Pragmatism	Reality is socially constructed— truth does not exist, but we can act on agreements, or "truth events"	Empirical science triangulatedwith qualitative methods	Peirce, Rorty
Critical Constructivism	Reality exists— truth claims made from a "partial perspective" are privileged over those made from a "god's-eye" perspective	Advocacy/action research	Haraway, Feenberg, Fischer
Social Constructivism, or Relativism	Reality is socially constructed— no truth claims are privileged over any other; all truth claims are power moves	Qualitative methods	Lincoln, Bijker

scientific method can or should distance us from the world. The concept of objectivity is, in her view, only a mask intended to obscure the political nature of scientific judgment; or, as Martin Heidegger has put it, the very idea that we can have a "worldview" at all is a Cartesian delusion unique to the culture of modernity (Heidegger 1977).

If, however, we reject positivistic science on the basis that it is radically reductive, does this suggest the opposite—that reality is socially constructed and that all claims to knowledge are equally true? This is the relativistic position of social constructivists placed at the bottom of table A.1. In the relativist worldview, because all truth claims are equal they should be understood as attempts to prop up one's own political position in the community. Haraway, however, rejects this position with equal enthusiasm as being only the flip side of positivism. Where positivists deny that they have any responsibility in what they study because they must maintain their distant objectivity, relativists deny their responsibility to act on the basis that all interpretations of reality are equally true. In neither case do investigators perceive having responsibility for what they observe.

Between these two extreme positions, and placed in the middle of table A.1, are two positions toward knowledge that I refer to alternately as critical constructivism and pragmatism. Both positions reject the spectator theory of knowledge held by positivists and relativists alike in favor of one that is transactional, or explicitly political (Hickman 2001). For Benjamin Barber, an exemplar of the pragmatist position, the issue of knowing the truth is not the slightest bit interesting (Barber 1984). Rather, he is more interested in relationships than facts and more interested in human practices than truth. This is to say that epistemology is political rather than prepolitical, as positivists would have it, or metaphysical, as relativists would have it. What we know is contingent but provides reasonable direction for action that is constantly corrected by what Barber (1984, 171) calls "public seeing"—an always emergent body of public knowledge.

The distinction between a pragmatist approach to knowing and that of critical constructivism that I am proposing in table A.1 is a fairly subtle one, and perhaps unnecessary. Larry Hickman (2001), for example, argues that although Andrew Feenberg comes from a critical theory background and refers to himself as a "critical constructivist," his critique of modern technology has evolved into one that is indistinguishable from John Dewey's pragmatism.[1] Nevertheless, as sympathetic as these attitudes toward knowledge may be, I include critical constructivism as a possibility in table A.1 primarily because it leads to the adoption of advocacy, or action research methods. For the moment, these two categories are adequate to describe the approach that I have taken in the study of Austin, Curitiba, and Frankfurt.

The primary implication for studying these cities through pragmatist or critical constructivist lenses was the need to "triangulate" potentially con-

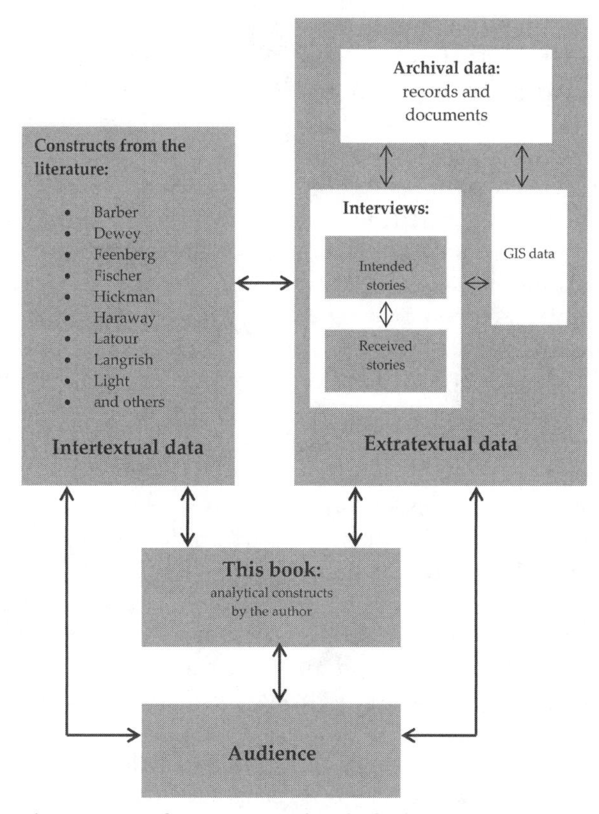

Figure A.1. The Hermeneutic-Diaglocic Method

flicting sources of data. The term *triangulation* was apparently originated by early radio operators. When an unknown signal was discovered, technicians were able to determine the geographic location of the signal by fixing two or more directional antennas upon it. Through simple plain geometric calculations, the operators were able to fix the signal in space. Social scientists have appropriated the term as a metaphor for credibility in inquiry. In this sense of constructing credible or triangulated meaning, I did not reveal the essential qualities of the cities studied by contemplating them from afar. Rather, interpretations were produced through a hermeneutic, dialogic process as illustrated in figure A.1.

In this study there are two levels of triangulation. The first balances intertextual data (the literatures), extratextual data (local primary sources), and my own analysis. The second layer of triangulation occurs within the extratextual data itself: First, archival data (records and documents) were gathered to reconstruct the official stories of each city. These primary sources were used to inform and verify the data gained through interviews with those who have produced the cases and those who have received them. Documents and interviews were then checked against quantitative data created by use of GIS, or geographic information systems. The use of GIS data has been helpful in two ways: first, the use of quantitative data tests the reliability of qualitative sources, and, second, GIS analysis is spatial so it checks the validity of historical data. I want to emphasize here, however, that I do not privilege either qualitative or quantitative data. Each is no more than one of many interpretive tools developed by distinct "epistemic communities" (Guy 2000).

Within the interpretation of what I have referred to as *extratextual data,* there is an implied symmetry in the value obtained from those with intentions toward the projects documented—its *producers*—and from those who have received the projects—their *receivers.* Both groups tell stories. This balance in the structure of inquiry is sympathetic to the advocates of reception theory—principally Hans Robert Jauss, the German literary critic—who maintains that one cannot understand a painting or a city solely through recuperating the intentions of authors, the study of the work's production, or through a description of its appearance. Rather, Jauss argues, works must be understood in terms of their production *and* reception (Holub 1984). This position reflects the general postmodern shift away from interest in the autonomous author and the work and toward the text and the reader. This dialogic method of interpretation has strongly influenced my reconstruction of what happened in Austin, Curitiba, and Frankfurt.

A principal purpose of this appendix is to make clear my own position in this inquiry is not that of a distanced scientist. Rather, both layers of triangulated data should be recognized as my own constructions. It is simply my hope that this triangulated structure lends the inquiry a degree of "trustworthiness" (Lincoln 1985) if not objectivity in the traditional sense of that term.

CHAPTER REFERENCES

Barber, B. (1984). *Strong Democracy: Participatory Politics for a New Age.* Berkeley: University of California Press.

Feenberg, A. (2003). "Pragmatism and critical theory of technology." *Techne* 7, 1 (Fall): 42–49.

Fischer, F. (2003). *Reframing public policy: Discursive politics and deliberative practices.* New York: Oxford University Press.

Guy, S. and E. Shove. (2000). *A sociology of energy, buildings, and the environment: Constructing knowledge, designing practice.* London: Routledge.

Haraway, D. (1995). "Situated knowledge: the science question in feminism and the privilege of partial perspective." Pp. 175–94 in A. Feenberg and A. Hannay, eds., *Technology and the politics of knowledge.* Bloomington: Indiana University Press.

Heidegger, M. (1977). "The age of the world picture" Pp. 115–54 in *The Question Concerning Technology and Other Essays, translated and with an introduction by William Lovitt*. New York: Harper and Rowe.
Hickman, L. A. (2001). *Philosophical tools for a technological culture: Putting pragmatism to work*. Bloomington: Indiana University Press.
Holub, R. C. (1984). *Reception theory: a critical introduction*. London: Methuen.
Lincoln, Y. and I. Guba. (1985). *Naturalistic inquiry*. Newbury Park, CA: Sage.

NOTE

1. For his part, Feenberg (Feenberg 2003) has objected to being considered a pragmatist, largely on the grounds that Dewey failed to take a critical position of capitalist technology as it unfolded in the twentieth century.

Index

About the Author

Steven A. Moore is the Bartlett Cocke Professor of Architecture and Planning at the University of Texas, Austin.